ENERGY
SUPPLIES, SUSTAINABILITY,
AND COSTS

ISSN 1534-1585

ENERGY
SUPPLIES, SUSTAINABILITY, AND COSTS

Sandra M. Alters

INFORMATION PLUS® REFERENCE SERIES
Formerly published by Information Plus, Wylie, Texas

THOMSON
GALE

Detroit • New York • San Francisco • San Diego • New Haven, Conn. • Waterville, Maine • London • Munich

Energy: Supplies, Sustainability, and Costs

Sandra M. Alters

Paula Kepos, Series Editor

Project Editor
John McCoy

Permissions
Margaret Abendroth, Edna Hedblad, Emma Hull

Composition and Electronic Prepress
Evi Seoud

Manufacturing
Drew Kalasky

LIBRARY OF CONGRESS CATALOGING-IN-PUBLICATION DATA

ISBN 0-7876-5103-6 (set)
ISBN 0-7876-9072-4
ISSN 1534-1585

Printed in the United States of America
10 9 8 7 6 5 4 3 2 1

TABLE OF CONTENTS

Energy is essential to human existence. The need for ever-increasing amounts of energy has resulted in some of the most profound social changes in human history. This chapter explores some of those changes, as well as current and future trends in energy production and consumption, public policy, and environmental concerns.

This chapter focuses on one of the planet's most important natural resources: oil. The discussion touches on many different aspects of the oil issue, including domestic and international production and consumption, price and demand trends, and environmental concerns stemming from oil spills.

Natural gas has become an important source of energy. This chapter covers the processes involved in producing, storing, and delivering natural gas; current trends in production and consumption; and predictions about the future of the gas industry.

Coal first rose to prominence as an energy source in the nineteenth century. After a period in which it was supplanted by oil as the world's primary energy source, coal regained some of its popularity during the oil crises of the 1970s and early 1980s. Besides tracing coal's rise and fall, and rise again, the chapter also discusses the different types and classifications of coal; mining methods; trends in price, production, and consumption; and the numerous environmental concerns that surround the use of this popular fossil fuel.

Nuclear energy offers significant advantages over fossil fuels but there are also significant downsides, including problems associated with the disposal of radioactive waste and the risk of nuclear-plant meltdowns, such as the Chernobyl disaster of 1986. The future of nuclear power is also discussed.

Renewable energy sources are those that replenish themselves naturally. Renewable energy sources, such as hydropower, solar and wind energy, geothermal power, and hydrogen, are all attractive alternatives to fossil-fuel sources, but they are not without their disadvantages. This chapter details some of these alternative sources, trends in their use, and the advantages and disadvantages of each.

Because oil, gas, coal, and uranium are nonrenewable resources—they are formed more slowly than they are consumed—it is important to know how much of each is recoverable, or potentially recoverable, from the earth. This chapter addresses that issue, focusing on domestic and international reserves of these resources; worldwide trends in the exploration for them; and the environmental impact of that exploration.

This chapter focuses on another major energy source: electricity. Among the topics discussed are domestic and international production and consumption of electricity; the attempt at deregulation of electric utilities in the United States; and projected trends in the domestic electric industry.

Public health is inextricably linked to the health of the environment, and the health of the environment is largely determined by how we use and dispose of our energy sources. This chapter analyzes this complex web, focusing on global warming, manufacturers' attempts to build more efficient automobiles and appliances, and projected trends in energy conservation.

PREFACE

Energy: Supplies, Sustainability, and Costs is part of the *Information Plus Reference Series.* The purpose of each volume of the series is to present the latest facts on a topic of pressing concern in modern American life. These topics include today's most controversial and most studied social issues: abortion, capital punishment, care for the elderly, crime, health care, energy, the environment, immigration, minorities, social welfare, women, youth, and many more. Although written especially for the high school and undergraduate student, this series is an excellent resource for anyone in need of factual information on current affairs.

By presenting the facts, it is Thomson Gale's intention to provide its readers with everything they need to reach an informed opinion on current issues. To that end, there is a particular emphasis in this series on the presentation of scientific studies, surveys, and statistics. These data are generally presented in the form of tables, charts, and other graphics placed within the text of each book. Every graphic is directly referred to and carefully explained in the text. The source of each graphic is presented within the graphic itself. The data used in these graphics are drawn from the most reputable and reliable sources, in particular from the various branches of the U.S. government and from major independent polling organizations. Every effort was made to secure the most recent information available. The reader should bear in mind that many major studies take years to conduct, and that additional years often pass before the data from these studies are made available to the public. Therefore, in many cases the most recent information available in 2005 dated from 2002 or 2003. Older statistics are sometimes presented as well, if they are of particular interest and no more-recent information exists.

Although statistics are a major focus of the *Information Plus Reference Series,* they are by no means its only content. Each book also presents the widely held positions and important ideas that shape how the book's subject is discussed in the United States. These positions are explained in detail and, where possible, in the words of those who support them. Some of the other material to be found in these books includes: historical background; descriptions of major events related to the subject; relevant laws and court cases; and examples of how these issues play out in American life. Some books also feature primary documents, or have pro and con debate sections giving the words and opinions of prominent Americans on both sides of a controversial topic. All material is presented in an even-handed and unbiased manner; the reader will never be encouraged to accept one view of an issue over another.

HOW TO USE THIS BOOK

The United States is the world's largest consumer of energy in all its forms. Gasoline and other fossil fuels power its cars, trucks, trains, and aircraft. Electricity generated by burning oil, coal, and natural gas—or from nuclear or hydroelectric plants—runs America's lights, telephones, televisions, computers, and appliances. Without a steady, affordable, and massive amount of energy, modern America could not exist. This book presents the latest information on U.S. energy consumption and production, compared with years past. Controversial issues such as the U.S. dependence on foreign oil; the possibility of exhausting fossil fuel supplies; and the harm done to the environment by mining, drilling, and pollution are explored.

Energy: Supplies, Sustainability, and Costs consists of nine chapters and three appendices. Each of the major elements of the U.S. energy system—such as coal, nuclear power, renewable energy sources, and electricity generation—has a chapter devoted to it. For a summary of the information covered in each chapter, please see the synopses provided in the Table of Contents at the front of the book. Chapters generally begin with an overview of the basic facts and background information on the chapter's topic, then proceed to examine sub-topics of particular interest. For example, Chapter 7: Energy Reserves—Oil, Gas, Coal, and Uranium begins with a description of the different ways in which natural resource reserves are measured and their reliability. The chapter then moves into an examination of proved U.S. reserves of oil and natural gas. Statistics on known and estimated reserves are presented, as are projections of how long these reserves will last if they continue to be consumed at current rates. This is followed by a discussion of the methods, costs, and consequences of oil and gas exploration. After this are sections on U.S. coal and uranium reserves. Then the chapter presents similar statistics on worldwide reserves of oil, coal, gas, and uranium. Readers can find their way through a chapter by looking for the section and sub-section headings, which are clearly set off from the text. Or, they can refer to the book's extensive index, if they already know what they are looking for.

Statistical Information

The tables and figures featured throughout *Energy: Supplies, Sustainability, and Costs* will be of particular use to the reader in learning about this topic. These tables and figures represent an extensive collection of the most recent and valuable statistics on energy production and consumption; for example, the amount of coal mined in the United States in a year, the rate at which energy consumption is increasing in the United States, and the percentage of U.S. energy that comes from renewable sources. Thomson Gale believes that making this information available to the reader is the most important way in which we fulfill the goal of this book: to help readers understand the topic of energy and reach their own conclusions about controversial issues related to energy use and conservation in the United States.

Each table or figure has a unique identifier appearing above it, for ease of identification and reference. Titles for the tables and figures explain their purpose. At the end of each table or figure, the original source of the data is provided.

In order to help readers understand these often complicated statistics, all tables and figures are explained in the text. References in the text direct the reader to the relevant statistics. Furthermore, the contents of all tables and figures are fully indexed. Please see the opening section of the index at the back of this volume for a description of how to find tables and figures within it.

Appendices

In addition to the main body text and images, *Energy: Supplies, Sustainability, and Costs* has three appendices. The first is the Important Names and Addresses directory. Here the reader will find contact information for a number of organizations that study energy. The second appendix is the Resources section, which is provided to assist the reader in conducting his or her own research. In this section, the author and editors of *Energy: Supplies, Sustainability, and Costs* describe some of the sources that were most useful during the compilation of this book. The final appendix is this book's index.

ADVISORY BOARD CONTRIBUTIONS

The staff of Information Plus would like to extend their heartfelt appreciation to the Information Plus Advisory Board. This dedicated group of media professionals provides feedback on the series on an ongoing basis. Their comments allow the editorial staff who work on the project to make the series better and more user-friendly. Our top priorities are to produce the highest-quality and most useful books possible, and the Advisory Board's contributions to this process are invaluable.

The members of the Information Plus Advisory Board are:

• Kathleen R. Bonn, Librarian, Newbury Park High School, Newbury Park, California

• Madelyn Garner, Librarian, San Jacinto College—North Campus, Houston, Texas

• Anne Oxenrider, Media Specialist, Dundee High School, Dundee, Michigan

• Charles R. Rodgers, Director of Libraries, Pasco-Hernando Community College, Dade City, Florida

• James N. Zitzelsberger, Library Media Department Chairman, Oshkosh West High School, Oshkosh, Wisconsin

COMMENTS AND SUGGESTIONS

The editors of the *Information Plus Reference Series* welcome your feedback on *Energy: Supplies, Sustainability, and Costs*. Please direct all correspondence to:

Editors
Information Plus Reference Series
27500 Drake Rd.
Farmington Hills, MI, 48331-3535

CHAPTER 1

AN ENERGY OVERVIEW

Energy is essential to life. Living creatures draw on energy flowing through the environment and convert it to forms they can use. The most fundamental energy flow for living creatures is the energy of sunlight, and the most important conversion is the act of primary production, in which plants and phytoplankton convert sunlight into biomass by photosynthesis. Earth's web of life, including human beings, rests on this foundation.

—*Energy in the United States: 1635–2000*, U.S. Energy Information Administration, 2001

A HISTORICAL PERSPECTIVE

Pre–Twentieth Century

People have always found ways to harness energy, such as using animals to do work or inventing machines to tap the power of wind or water. The industrialization of the modern world starting in the eighteenth century was accompanied by the widespread use of such fossil fuels as coal, oil, and natural gas.

Significant use and management of energy resulted in one of the most profound social changes in history within a few generations. In the early 1800s most Americans lived in rural areas and worked in agriculture. The country ran mainly on wood fuel. One hundred years later, most Americans were city dwellers and worked in industry. America had become the world's largest producer and consumer of fossil fuels, had roughly tripled its per capita use of energy, and had become a global superpower.

The United States has always been a resource-abundant nation. But it was not until the Industrial Revolution in the mid-1800s that the total work output of engines surpassed that of work animals. As the United States industrialized, coal began to replace wood as a primary fuel. Then, as industrialization proceeded, petroleum and natural gas replaced coal for many applications. The United States has since relied heavily on these three fossil fuels—coal, petroleum, and natural gas.

FIGURE 1.1

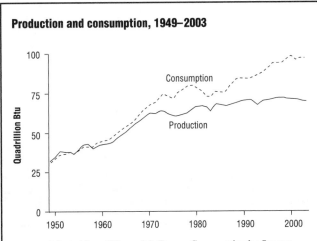

Production and consumption, 1949–2003

SOURCE: Adapted from "Figure 1.3. Energy Consumption by Source: Production and Consumption, 1949–2003," in *Annual Energy Review 2003*, U.S. Department of Energy, Energy Information Administration, Office of Energy Markets and End Use, September 7, 2004, http://www.eia.doe.gov/emeu/aer/pdf/aer.pdf (accessed September 28, 2004)

The Twentieth Century

For much of its history the United States has been nearly energy self-sufficient, although small amounts of coal were imported from Britain in colonial times. Through the 1950s domestic energy production and consumption were nearly equal. During the 1960s consumption slightly outpaced production. By the 1970s the gap had widened, and by 2003 this gap was quite significant. (See Figure 1.1.) Since the 1970s energy imports have been used to try to close the energy production/consumption gap. However, America's dependence on other countries for some of its energy needs has brought with it significant problems.

OIL CRISIS IN THE 1970s. In 1973 the United States supported Israel in the Yom Kippur War, which Israel fought against its neighboring Arab countries. In response, several of these Arab nations cut off exports of oil to the

FIGURE 1.2

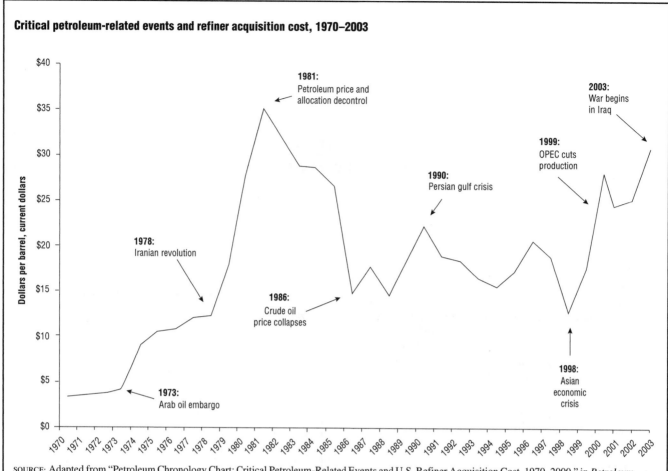

Critical petroleum-related events and refiner acquisition cost, 1970–2003

SOURCE: Adapted from "Petroleum Chronology Chart: Critical Petroleum-Related Events and U.S. Refiner Acquisition Cost, 1970–2000," in *Petroleum Chronology of Events 1970–2000,* U.S. Department of Energy, Energy Information Administration, May 2002, http://www.eia.doe.gov/pub/oil_gas/petroleum/analysis_publications/chronology/petroleumchronology2000.htm (accessed November 23, 2004)

United States and decreased exports to the rest of the world. The U.S. embargo was lifted six months later, but the price of oil tripled from the 1973 average to about $12 per barrel. (See Figure 1.2.) Not only did Americans (and others around the world) face sudden price hikes for products produced from oil, such as gasoline and home heating oil, but they faced temporary shortages as well. The energy problem quickly became an energy crisis, which led to occasional blackouts in cities and industries, temporary shutdowns of factories and schools, and frequent lines at gasoline service stations. The sudden increase in energy prices in the early 1970s is widely considered to have been a major cause of the economic recession of 1974 and 1975.

Oil prices increased even more in the late 1970s. The Iranian revolution began in late 1978 and resulted in a significant drop in Iranian oil production from 1978 to 1981. During this same period the Iran-Iraq war began, and many other Persian Gulf countries decreased their output as well. Companies and governments began to stockpile oil. As a result, prices continued to rise and reached a peak in 1981. (See Figure 1.2.)

OIL PRICES FALL IN THE 1980s. In early 1981 the U.S. government responded to the oil crisis by removing price and allocation controls on the oil industry. That is, the government no longer controlled domestic crude oil prices or restricted exports of petroleum products, preferring to allow the marketplace and competition to determine the price of crude oil. Therefore, domestic oil prices rose to meet foreign oil prices.

As a result of these increasingly high prices, individuals and industry stepped up their conservation efforts and switched to alternative fuels. The demand for crude oil declined. But the Organization of Petroleum Exporting Countries (OPEC), and particularly Saudi Arabia, cut its output in the first half of the 1980s to keep the price from declining dramatically. (In late 2004 member countries of OPEC were Algeria, Indonesia, Iran, Iraq, Kuwait, Libya, Nigeria, Qatar, Saudi Arabia, United Arab Emirates, and Venezuela.)

In 1985 Saudi Arabia moved to increase its market share of crude oil exports by increasing its production. (Saudi Arabia was and still is the world's largest producer and exporter of oil, and is a key member of OPEC.) Other OPEC members followed suit, which resulted in a glut of crude oil on the world market. Crude oil prices fell sharply in early 1986, and imports increased.

THE UPS AND DOWNS OF OIL PRICES IN THE 1990s AND EARLY 2000s. In August 1990 Iraq invaded Kuwait, and the public feared an oil shortfall caused by the United Nations (UN) embargo on all crude oil and oil products from both countries. Prices rose suddenly and sharply, but non-OPEC countries in Central America, western Europe, and the Far East, along with the United States, stepped up their production to fill the gap in world supplies. After the UN approved the use of force against Iraq during the Persian Gulf crisis in October 1990, prices fell quickly. (See Figure 1.2.)

The collapse of Asian economies in the mid-1990s led to a further drop in the demand for energy, and petroleum prices dipped sharply in the late 1990s. OPEC reacted by curtailing production, which boosted prices in 2000. (See Figure 1.2.) World crude oil prices then declined through 2001 as global demand dropped because of weakening economies (especially in the United States) and a drop in jet fuel demand following the September 11, 2001, terrorist attacks.

In late 2002 terrorist attacks and counterattacks between Israelis and Palestinians caused concerns that Iraq might halt its crude oil shipments to countries that supported the Jewish state of Israel over Islamic Palestine. Additionally, concerns existed that the Middle East region might become destabilized should the United States invade Iraq, which has the second-largest oil reserve in the world. Moreover, Venezuelan oil workers went on strike, which cut off Venezuelan oil imports. These three factors were the primary causes of the rise in crude oil prices by the end of 2002.

VOLATILITY AND RECORD HIGHS IN OIL PRICES IN 2003 AND 2004. In early 2003 war with Iraq seemed imminent. In addition, due to a cold winter and the Venezuelan strike, U.S. commercial crude oil inventories had declined. As a result, crude oil prices rose to a twenty-nine-month high—nearly $40 a barrel in February 2003. On March 19, 2003, the war in Iraq began, and Iraqi oil fields were shut down. Other oil-producing countries stepped up production to offset the shortfall, however. In addition, the Venezuelan strike had ended. As a result, oil prices declined dramatically to about $27 a barrel by the beginning of May 2003.

But lower prices did not prevail. By June 2003 the price of oil rose above $30 a barrel because supplies of crude were low at the start of the summer driving season. The price of crude oil continued to climb over the summer to slightly more than $31 a barrel due to sustained political turmoil in Iraq. Then, in the fall of 2003, the U.S. dollar sank to a record low against the euro, and government data indicated that crude inventories were continuing to

dwindle. By December 2003 crude oil prices had risen to nearly $34 a barrel.

Events in 2004 brought no relief to the rise in crude oil prices, which continued to be affected by political uncertainty, the weakened U.S. dollar, and tight supplies. Increases in crude oil prices in 2004 also reflected a growing demand from the world's three largest oil consumers: the United States, China, and Japan. In addition, oil prices rose as concerns about terrorism in Spain, Iraq, Pakistan, Saudi Arabia, and other troubled areas increased. Moreover, sabotage of Iraq's northern oil pipelines prevented the country from producing the oil that was expected. By March 2004 oil prices soared to a thirteen-year high of about $38 a barrel. By August the price of crude reached more than $45 a barrel, and by October oil closed above $50 a barrel for the first time.

GOVERNMENTAL ENERGY POLICIES

Under President Ronald Reagan

In 1977 President Jimmy Carter, a Democrat, described the energy problem of the time as one that could only be "effectively addressed by a Government that accepts responsibility for dealing with it comprehensively and by a public that understands its seriousness and is ready to make necessary sacrifices." When Republican Ronald Reagan took over the presidency, however, he downplayed the importance of government responsibility for dealing with the energy problem. The Reagan administration sharply cut federal programs for energy and opposed government intervention in energy markets. For example, the administration refused to tax energy imports, which may have stimulated domestic production and conservation. Reagan believed that the expansion of the federal government's role in energy policy was counterproductive and misguided. His administration transferred the center of the decision-making process to the states, the private sector, and individuals.

Under President George H. W. Bush

The subsequent Republican administration of President George H. W. Bush continued the Reagan policy of limiting government regulation of the energy industry. In 1991 President Bush unveiled a long-awaited energy policy that promised to reduce U.S. dependence on foreign oil by increasing domestic oil production and the use of nuclear power. Bush's aim was to rely on "the power of the marketplace, the common sense of the American people, and the responsible leadership of government and industry." He planned to achieve this by, among other proposals, producing additional oil from environmentally sensitive areas, encouraging pipeline construction, simplifying the construction permit process for nuclear power plants, and increasing competition in the production of electricity. His proposals did not include government-directed conservation efforts or tax incentives.

Conservationists disagreed with President Bush, objecting to increased offshore drilling, especially in the coastal plain of the Arctic National Wildlife Refuge (ANWR) in Alaska. They also wanted to see automobile fuel economy improved and conservation methods stressed, rather than the use of nuclear power.

Under President Bill Clinton

The Democratic administration of President Bill Clinton brought government involvement back into energy and environmental issues, although the major energy bills proposed by this administration were not passed into law by the Republican-dominated Congress of the time. Nevertheless, the Clinton administration did increase funds for alternative energy research, mandate new energy efficiency measures, and enforce emission standards. The administration also opened up several areas for oil exploration, including some Alaskan and offshore areas.

Under President George W. Bush

The major energy policy goals of the first term of Republican president George W. Bush were to increase and diversify the sources of America's oil supplies and to make energy security a priority. The Bush administration encouraged efforts to import more Russian crude oil and reopened the American embassy in Equatorial Guinea, an oil-rich nation.

DOMESTIC ENERGY USAGE

Domestic Production

The total domestic energy production of the United States—the amount of fossil fuels and other forms of energy that was mined, pumped, or otherwise originated in the United States—has more than doubled since 1949, rising from 31.7 quadrillion Btu (British thermal units) in 1949 to 70.5 quadrillion Btu in 2003. (See Table 1.1 and Figure 1.3.) One quadrillion Btu equals the energy produced by approximately 170 million barrels of crude oil. Large production and consumption figures are given in these units to make it easier to compare the various types of energy, which come in different forms.

Table 1.1 and Figure 1.4 show that from 1949 to 2003 the energy produced in the United States from coal generally increased steadily. The production of oil rose until 1972 but by 1995 had declined to about the levels produced in the early 1950s, which were in the 13 quadrillion Btu range (not shown in Table 1.1.) The decline continued through 2003. Likewise, natural gas production rose until 1972. It then declined from 1973 through 1986. Since 1986 natural gas production has risen to levels similar to those in the late 1970s. The energy produced from nuclear power has increased over the past fifty-four years, while the energy produced from hydroelectric and biofuel power (wood, waste, alcohol) in recent decades has

FIGURE 1.3

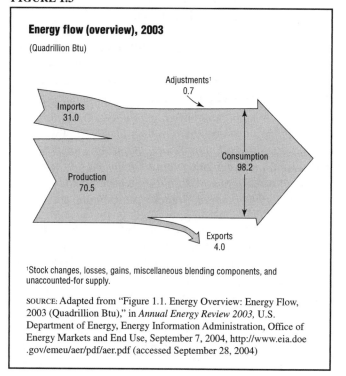

Energy flow (overview), 2003

(Quadrillion Btu)

Adjustments[1]
0.7

Imports
31.0

Consumption
98.2

Production
70.5

Exports
4.0

[1]Stock changes, losses, gains, miscellaneous blending components, and unaccounted-for supply.

SOURCE: Adapted from "Figure 1.1. Energy Overview: Energy Flow, 2003 (Quadrillion Btu)," in *Annual Energy Review 2003*, U.S. Department of Energy, Energy Information Administration, Office of Energy Markets and End Use, September 7, 2004, http://www.eia.doe.gov/emeu/aer/pdf/aer.pdf (accessed September 28, 2004)

remained relatively steady. More coal was produced in the United States during 2003 than any other energy source, with natural gas in second place. Oil was the third largest form of energy, while nuclear electric power was fourth.

Domestic Consumption

While total domestic energy production has more than doubled since 1949, the total domestic energy consumption—the amount of fossil fuels and other forms of energy used by the people and industry of the United States—has more than tripled, rising from 30 quadrillion Btu in 1949 to 98.2 quadrillion Btu in 2003. (See Figure 1.1 and Figure 1.3.) Domestic energy consumption more than doubled from 1949 to 1973, increasing from 30 to 74 quadrillion Btu. Meanwhile, the economy grew at about the same rate, so the increased consumption of energy reflected the growth in the economy. That is, as the nation grew, it used more fuel—mainly more petroleum and natural gas.

But after the huge 1973 oil price increases, energy consumption fell, rose, and fell again, eventually returning to 1973 levels by 1986. (See Figure 1.1.) Following the drop in crude oil prices in 1986, U.S. imports of oil began to rise, and energy consumption increased, reaching an all-time high of 98.9 quadrillion Btu in 2000. According to U.S. Census Bureau data, the U.S. population grew by 88% from 1950 to 2000, while energy consumption rose by 181% during the same period. After 2000 energy consumption leveled off somewhat. By 2003 it had dropped slightly, to 98.2 Btu. (See Figure 1.3)

Before the 1973 oil crisis, U.S. energy consumption increased quickly. (See Figure 1.1.) After 1973 energy con-

TABLE 1.1

Energy production by source, selected years, 1949–2003

(Quadrillion Btu)

Year	Coal	Natural gas (dry)	Crude oil[2]	Natural gas plant liquids	Total	Nuclear electric power	Hydroelectric pumped storage[3]	Conventional hydroelectric power	Wood waste alcohol[4]	Geothermal	Solar	Wind	Total	Total
			Fossil fuels					Renewable energy[1]						
1949	11.974	5.377	10.683	0.714	28.748	0.000	(s)	1.425	1.549	NA	NA	NA	2.974	31.722
1950	14.060	6.233	11.447	0.823	32.563	0.000	(s)	1.415	1.562	NA	NA	NA	2.978	35.540
1955	12.370	9.345	14.410	1.240	37.364	0.000	(s)	1.360	1.424	NA	NA	NA	2.784	40.148
1960	10.817	12.656	14.935	1.461	39.869	0.006	(s)	1.608	1.320	0.001	NA	NA	2.929	42.804
1965	13.055	15.775	16.521	1.883	47.235	0.043	(s)	2.059	1.335	0.004	NA	NA	3.398	50.676
1970	14.607	21.666	20.401	2.512	59.186	0.239	(s)	2.634	1.431	0.011	NA	NA	4.076	63.501
1971	13.186	22.280	20.033	2.544	58.042	0.413	(s)	2.824	1.432	0.012	NA	NA	4.268	63.723
1972	14.092	22.208	20.041	2.598	58.938	0.584	(s)	2.864	1.503	0.031	NA	NA	4.398	63.920
1973	13.992	22.187	19.493	2.569	58.241	0.910	(s)	2.861	1.529	0.043	NA	NA	4.433	63.585
1974	14.074	21.210	18.575	2.471	56.331	1.272	(s)	3.177	1.540	0.053	NA	NA	4.769	62.372
1975	14.989	19.640	17.729	2.374	54.733	1.900	(s)	3.155	1.499	0.070	NA	NA	4.723	61.357
1976	15.654	19.480	17.262	2.327	54.723	2.111	(s)	2.976	1.713	0.078	NA	NA	4.768	61.602
1977	15.755	19.565	17.454	2.327	55.101	2.702	(s)	2.333	1.838	0.077	NA	NA	4.249	62.052
1978	14.910	19.485	18.434	2.245	55.074	3.024	(s)	2.937	2.038	0.064	NA	NA	5.039	63.137
1979	17.540	20.076	18.104	2.286	58.006	2.776	(s)	2.931	2.152	0.084	NA	NA	5.166	65.948
1980	18.598	19.908	18.249	2.254	59.008	2.739	(s)	2.900	2.485	0.110	NA	NA	5.494	67.241
1981	18.377	19.699	18.146	2.307	58.529	3.008	(s)	2.758	2.590	0.123	NA	NA	5.471	67.007
1982	18.639	18.319	18.309	2.191	57.458	3.131	(s)	3.266	2.615	0.105	(s)	NA	5.985	66.574
1983	17.247	16.593	18.392	2.184	54.416	3.203	(s)	3.527	2.831	0.129	(s)	(s)	6.488	64.106
1984	19.719	18.008	18.848	2.274	58.849	3.553	(s)	3.386	2.880	0.165	(s)	(s)	6.431	68.832
1985	19.325	16.980	18.992	2.241	57.539	4.076	(s)	2.970	2.864	0.198	(s)	(s)	6.033	67.647
1986	19.509	16.541	18.376	2.149	56.575	4.380	(s)	3.071	2.841	0.219	(s)	(s)	6.132	67.087
1987	20.141	17.136	17.675	2.215	57.167	4.754	(s)	2.635	2.823	0.229	(s)	(s)	5.687	67.608
1988	20.738	17.599	19.279	2.260	57.875	5.587	(s)	2.334	2.937	0.217	(s)	(s)	5.489	68.951
1989	21.346	17.847	16.117	2.158	57.468	5.602	(s)	2.837	3.062	0.317	0.055	0.022	6.294	69.364
1990	22.456	18.326	15.571	2.175	58.529	6.104	-0.036	3.046	2.662	0.336	0.060	0.029	6.133	70.729
1991	21.594	18.229	15.701	2.306	57.829	6.422	-0.047	3.016	2.702	0.346	0.063	0.031	6.158	70.362
1992	21.629	18.375	15.223	2.363	57.590	6.479	-0.043	2.617	2.847	0.349	0.064	0.030	5.907	69.933
1993	20.249	18.584	14.494	2.408	55.736	6.410	-0.042	2.892	R2.803	0.364	0.066	0.031	R6.156	R68.260
1994	22.111	19.348	14.103	2.391	57.952	6.694	-0.035	2.683	2.939	0.338	0.069	0.036	6.065	70.676
1995	22.029	19.082	13.887	2.442	57.440	7.075	-0.028	3.205	3.068	0.294	0.070	0.033	6.669	71.156
1996	22.684	19.344	13.723	2.530	58.281	7.087	-0.032	3.590	3.127	0.316	0.070	0.033	7.137	72.472
1997	23.211	19.394	13.658	2.495	58.758	6.597	-0.041	3.640	3.006	0.325	0.071	0.034	7.075	72.389
1998	23.935	19.613	13.235	2.420	59.204	7.068	-0.046	3.297	2.835	0.328	0.070	0.031	6.561	72.787
1999	23.186	19.341	12.451	2.528	57.505	7.610	-0.062	3.268	2.885	0.331	0.069	0.046	6.599	71.652

TABLE 1.1

Energy production by source, selected years, 1949–2003 [CONTINUED]

(Quadrillion Btu)

Year	Fossil fuels					Nuclear electric power	Renewable energy[1]							Total
	Coal	Natural gas (dry)	Crude oil[2]	Natural gas plant liquids	Total		Hydro-electric pumped storage[3]	Conventional hydroelectric power	Wood waste alcohol[4]	Geothermal	Solar	Wind	Total	
2000	22.623	19.662	12.358	2.611	57.254	7.862	−0.057	2.811	2.907	0.317	0.066	0.057	6.158	71.218
2001	R23.529	R20.205	12.282	2.547	R58.563	R8.033	−0.090	2.201	R2.640	0.311	0.065	0.068	R5.286	R71.792
2002	R22.698	R19.495	R12.163	R2.559	R56.915	R9.143	RP−0.088	RP2.675	R2.791	RP0.328	P0.064	RP0.105	RP5.963	R70.933
2003	P22.311	P19.641	P12.145	P2.343	P56.440	P7.973	P−0.088	P2.779	P2.884	P0.314	P0.063	P0.108	P6.150	P70.474

[1]Electricity net generation from conventional hydroelectric power, geothermal, solar, and wind; consumption of wood, waste, and acohol fuels; geothermal heat pump and direct use energy; and solar thermal direct use energy.

[2]Includes lease condensate.

[3]Pumped storage facility production minus energy used for pumping.

[4]"Alcohol" is ethanol blended into motor gasoline.

[5]Included in "Conventional Hydroelectric Power."

R = Revised.

P = Preliminary.

NA=Not available.

(s) = Less than 0.0005 quadrillion btu.

Note: Totals may not equal sum of components due to independent rounding.

SOURCE: "Energy Production by Source, Selected Years, 1949–2003, (Quadrillion Btu), " in *Annual Energy Review 2003*, U.S. Department of Energy, Energy Information Administration, Office of Energy Markets and Energy Use, September 7, 2004, http://www.eia.doe.gov/emeu/aer/pdf/aer.pdf (accessed September 28, 2004)

FIGURE 1.4

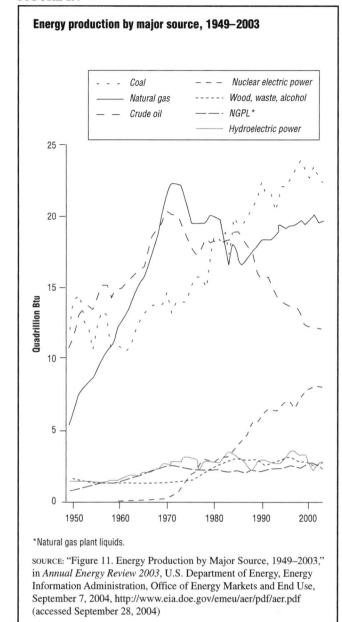

Energy production by major source, 1949–2003

- - - Coal
—— Natural gas
— — Crude oil
— — Nuclear electric power
- - - - Wood, waste, alcohol
—·— NGPL*
········ Hydroelectric power

*Natural gas plant liquids.

SOURCE: "Figure 11. Energy Production by Major Source, 1949–2003," in *Annual Energy Review 2003*, U.S. Department of Energy, Energy Information Administration, Office of Energy Markets and End Use, September 7, 2004, http://www.eia.doe.gov/emeu/aer/pdf/aer.pdf (accessed September 28, 2004)

sumption continued to increase but less sharply, as Americans became more efficient and used less energy to accomplish more. Energy consumption shifted slightly away from petroleum and natural gas toward electricity generated by other fuels. In 1973 petroleum and natural gas accounted for 77% of total energy consumption; by 2003 their share had dropped to 63% (40% petroleum and 23% natural gas).

Figure 1.5 shows energy production and consumption flows, including types of energy sources, in 2003. Coal, which in 1973 accounted for 17% of all energy consumed, accounted for 23.1% in 2003, or 22.7 quadrillion Btu out of a total of 98.2 quadrillion Btu. Nuclear power, which contributed barely 1% of the nation's consumption in 1973, accounted for 8.1% in 2003. Renewable energy sources (hydroelectric, solar, biofuels, and wind energy) accounted for 6.3% of energy consumed.

ENERGY IMPORTS AND EXPORTS

After 1958 the United States consumed more energy than it produced (see Figure 1.1) but made up the difference by importing energy. Imports (mainly oil) grew rapidly from 1953 through 1973 as the U.S. economy grew with the use of inexpensive oil. In 1973 net imports of petroleum reached almost 13 quadrillion Btu.

Although the Arab oil embargo of 1973–74—coupled with increased oil prices—momentarily slowed growth in petroleum imports, the general increase continued, with imports exceeding 18 quadrillion Btu in 1977. That year, U.S. dependence on petroleum imports rose to 46.5% of the nation's oil consumption. Despite the lesson of 1973, it took a second round of price increases in 1979–80, as shown in Figure 1.2, accompanied by lengthy and frustrating lines at gas stations, to convince Americans that they had to become less dependent on imported oil, conserve resources more, or both. By 1985 U.S. dependence on foreign oil had decreased sharply, to 27.3% of oil consumption. (See Figure 1.6.)

After 1985 the U.S. dependence on foreign sources of oil started to gradually increase, as a drop in crude oil prices drove up demand. When Iraq invaded Kuwait in 1990, the potential threat to the flow of oil to America and other industrialized nations was one of the reasons the United States challenged Saddam Hussein and eventually declared war at that time. After the terrorist attacks of September 11, 2001, and the subsequent Bush administration "War on Terror," the concept of energy independence, or at least less energy dependence, became increasingly important. In 2002 imports fell slightly for the first time since 1995. But in 2003 the United States was at war again with Iraq, and imported oil accounted for a record 56.1% of U.S. oil consumption. (See Figure 1.6.) That year net imports of petroleum reached 24.1 quadrillion Btu. (see Table 1.2)

Although the United States imports energy in the form of oil, it exports energy in the form of coal. Since 1950 America has produced more coal than it has consumed and has been an exporter of coal to other nations. In 2003 coal exports totaled 1.1 quadrillion Btu, nearly 28% of U.S. energy exports. (See Figure 1.7 and Table 1.2.)

FOSSIL FUEL PRODUCTION PRICES

Production prices are the value of fuel produced. The combined production prices of fossil fuels (crude oil, natural gas, and coal) slowly declined from 1949 through 1972. (See Figure 1.8 and Table 1.3.) These prices then increased dramatically from 1973 through 1981, and fell through 1998. To indicate how marked this decline in fossil fuel prices was, the composite value of all fossil fuel prices (in real dollars, which account for inflation) dropped by more than two-thirds from 1981 to 1998, from $4.64 per million Btu to $1.46 per million Btu. These huge

FIGURE 1.5

Energy flow (detail), 2003

[Quadrillion Btu]

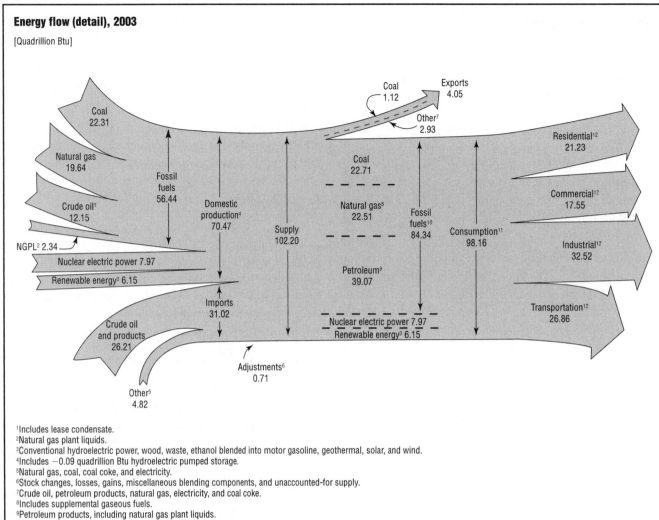

[1]Includes lease condensate.
[2]Natural gas plant liquids.
[3]Conventional hydroelectric power, wood, waste, ethanol blended into motor gasoline, geothermal, solar, and wind.
[4]Includes −0.09 quadrillion Btu hydroelectric pumped storage.
[5]Natural gas, coal, coal coke, and electricity.
[6]Stock changes, losses, gains, miscellaneous blending components, and unaccounted-for supply.
[7]Crude oil, petroleum products, natural gas, electricity, and coal coke.
[8]Includes supplemental gaseous fuels.
[9]Petroleum products, including natural gas plant liquids.
[10]Includes 0.05 quadrillion btu of coal coke net imports.
[11]Includes, in quadrillion btu, −0.09, hydroelectric pumped storage; −0.24 ethanol blended into motor gasoline, which is accounted for in both fossil fuels and renewable energy but counted only once in total consumption; and 0.02 electricity net imports.
[12]Primary consumption, electricity retail sales, and electrical system energy losses, which are allocated to the end-use sectors in proportion to each sector's share of total electricity retail sales.
Notes: Data are preliminary. Totals may not equal sum of components due to independent rounding.

SOURCE: "Diagram 1. Energy Flow, 2003 (Quadrillion Btu)," in *Annual Energy Review 2003*, U.S. Department of Energy, Energy Information Administration, Office of Energy Markets and End Use, September 7, 2004, http://www.eia.doe.gov/emeu/aer/pdf/aer.pdf (accessed September 28, 2004)

drops meant economic problems in fuel-producing American states such as Texas, Louisiana, Oklahoma, Montana, West Virginia, and Ohio, and in energy-exporting countries such as many Middle Eastern nations, Nigeria, Indonesia, Venezuela, and Trinidad. On the other hand, they were a windfall for industries that used a lot of energy, such as airlines, trucking companies, steel mills, and electric utilities. Since 1998 fossil fuel production prices have been rising, reaching $2.95 (in real dollars) per million Btu in 2003.

The production prices of both crude oil (the most expensive of the fossil fuels) and natural gas followed a pattern of rising and falling similar to that of the fossil fuel composite price from 1949 to 2000. (See Figure 1.8 and Table 1.3.) After slowly declining from 1949 through 1972, crude oil production prices rose the most dramatically of all the fossil fuels from 1973 to 1981, topping out at $9.27 per million Btu in 1981. The price then tumbled to $1.94 in 1998. But it then rose sharply during the next two years, reaching $4.61 in 2000. After a bit of a decline in 2001 and 2002, the crude oil production price rose to $4.50 in 2003. For natural gas, the price sank from $3.56 per million Btu in 1983 to $1.83 in 1998. It jumped up to an all-time high of $4.26 by 2003.

The story of coal prices is a bit different from that of the other fossil fuels. Coal production prices rose from 1970 to 1975, but then declined steadily through 2000 (not rising from 1975 to 1981 as did natural gas and crude oil production prices) and leveled out through 2003. (See Figure 1.8 and Table 1.3.) Its peak price in 1975 was $2.22 per million Btu in real dollars, and by 2000 it dropped to an all-time low of 80 cents per million Btu. In 2003 coal production prices were 82 cents per million Btu in real dollars.

FIGURE 1.6

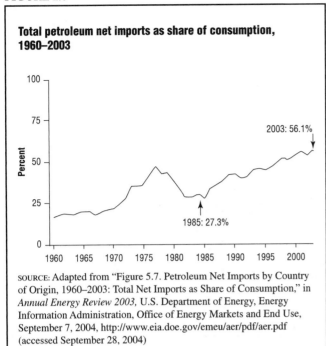

Total petroleum net imports as share of consumption, 1960–2003

SOURCE: Adapted from "Figure 5.7. Petroleum Net Imports by Country of Origin, 1960–2003: Total Net Imports as Share of Consumption," in *Annual Energy Review 2003,* U.S. Department of Energy, Energy Information Administration, Office of Energy Markets and End Use, September 7, 2004, http://www.eia.doe.gov/emeu/aer/pdf/aer.pdf (accessed September 28, 2004)

FIGURE 1.7

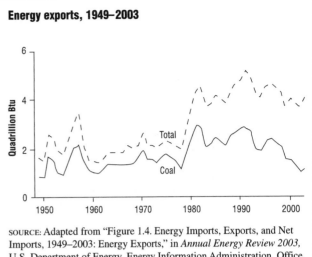

Energy exports, 1949–2003

SOURCE: Adapted from "Figure 1.4. Energy Imports, Exports, and Net Imports, 1949–2003: Energy Exports," in *Annual Energy Review 2003,* U.S. Department of Energy, Energy Information Administration, Office of Energy Markets and End Use, September 7, 2004, http://www.eia.doe.gov/emeu/aer/pdf/aer.pdf (accessed September 28, 2004)

ENERGY USE BY SECTOR

Energy use can be classified into four main "end-use" sectors: residential, commercial, industrial, and transportation. Historically, industry has been the largest energy-consuming sector of the economy, followed by the transportation, residential, and commercial sectors, in that order. In 2003 industry used about 33 quadrillion Btu, compared to approximately 27 quadrillion Btu in the transportation sector, 21 quadrillion Btu in the residential sector, and 18 quadrillion Btu in the commercial sector. (See Figure 1.9.)

Within sectors, energy sources have changed over time. For example, in the commercial sector, coal was the leading energy source through 1953 but declined dramatically in favor of petroleum (through 1962) and then natural gas (since 1963). In 1990 energy in the form of electricity pulled ahead as the leading energy source. (See Figure 1.10.) Similarly, coal was the leading energy source in the residential sector in 1949. (See Figure 1.11.) Natural gas quickly took over, with petroleum in second place. In 1979 electricity took over second place from petroleum. Industry used more coal than natural gas or petroleum through 1957, but after that natural gas and petroleum took over as nearly equally preferred energy sources. (See Figure 1.12.) In transportation, reliance on petroleum has been increasing since 1949. (See Figure 1.13.)

Not included in the four main sectors of energy consumption is the electric power sector. This sector includes electric utilities that generate, transmit, distribute, and sell electricity for use by the public. The electric power generated by this sector is then consumed by the other four sec-

tors, as noted in Figure 1.10, Figure 1.11, Figure 1.12, and Figure 1.13 and in the previous paragraph. To generate electricity that comes into homes, businesses, and industry, the electricity sector must harness energy from renewable sources such as the sun, the wind, water movement, and geothermal sources, or it must generate electricity from nuclear power or the burning of fossil fuels. Figure 1.14 shows that the electric power sector generates most of the electricity for the United States by burning coal. The electric power sector consumed approximately 20 quadrillion Btu of coal in 2003, a figure that has increased tenfold since 1949 and accounts for almost all coal usage in the United States. Nuclear power is the next most used fuel, followed by natural gas, renewable energy sources (hydroelectric, wood, waste, geothermal, solar, and wind), and petroleum. As Figure 1.14 shows, renewable sources are used very little to generate the nation's electricity.

INTERNATIONAL ENERGY USAGE

World Production

World production of primary energy rose from 215 quadrillion Btu in 1970 to 405 quadrillion Btu in 2002. (See Table 1.4.) The Energy Information Administration (EIA) of the U.S. Department of Energy stated in its 2004 report *Annual Energy Review 2003* that the world's total output of primary energy increased by 88% from 1970 to 2002. In 2002 fossil fuels were the most heavily produced fuel, accounting for 85% of all energy produced worldwide. Renewable energy accounted for 8% of all energy produced worldwide that year, and nuclear power accounted for 7%.

In 2002 the United States, Russia, and China were by far the leading producers of energy, followed by Saudi Arabia, Canada, and the United Kingdom. (See Figure 1.15.) According to the EIA report *Annual Energy Review*

TABLE 1.2

Energy imports, exports, and net imports, 1949–2003

(Quadrillion Btu)

Year	Imports					Exports					Net imports				
	Coal	Natural gas	Petroleum[1]	Other[2]	Total	Coal	Natural gas	Petroleum	Other[2]	Total	Coal	Natural gas	Petroleum[1]	Other[2]	Total
1949	0.01	0.00	1.43	0.01	1.45	0.88	0.02	0.68	0.01	1.59	−0.87	−0.02	0.75	(s)	−0.14
1950	0.01	0.00	1.89	0.02	1.91	0.79	0.03	0.64	0.01	1.47	−0.78	−0.03	1.24	0.01	0.45
1955	0.01	0.01	2.75	0.02	2.79	1.46	0.03	0.77	0.01	2.29	−1.46	−0.02	1.98	(s)	0.50
1960	0.01	0.16	4.00	0.02	4.19	1.02	0.01	0.43	0.01	1.48	−1.02	0.15	3.57	0.01	2.71
1965	(s)	0.47	5.40	0.01	5.89	1.38	0.03	0.39	0.03	1.83	−1.37	0.44	5.01	−0.02	4.06
1970	(s)	0.85	7.47	0.02	8.34	1.94	0.07	0.55	0.08	2.63	−1.93	0.77	6.92	−0.05	5.71
1971	(s)	0.96	8.54	0.03	9.53	1.55	0.08	0.47	0.05	2.15	−1.54	0.88	8.07	−0.02	7.38
1972	(s)	1.05	10.30	0.04	11.39	1.53	0.08	0.47	0.04	2.12	−1.53	0.97	9.83	(s)	9.27
1973	(s)	1.06	13.47	0.08	14.61	1.43	0.08	0.49	0.04	2.03	−1.42	0.98	12.98	0.04	12.58
1974	0.05	0.99	13.13	0.14	14.30	1.62	0.08	0.46	0.04	2.20	−1.57	0.91	12.66	0.10	12.10
1975	0.02	0.98	12.95	0.08	14.03	1.76	0.07	0.44	0.05	2.32	−1.74	0.90	12.51	0.03	11.71
1976	0.03	0.99	15.67	0.07	16.76	1.60	0.07	0.47	0.04	2.17	−1.57	0.92	15.20	0.03	14.59
1977	0.04	1.04	18.76	0.11	19.95	1.44	0.06	0.51	0.04	2.05	−1.40	0.98	18.24	0.07	17.90
1978	0.07	0.99	17.82	0.21	19.11	1.08	0.05	0.77	0.02	1.92	−1.00	0.94	17.06	0.19	17.19
1979	0.05	1.30	17.93	0.18	19.46	1.75	0.06	1.00	0.04	2.86	−1.70	1.24	16.93	0.13	16.60
1980	0.03	1.01	14.66	0.10	15.80	2.42	0.05	1.16	0.07	3.69	−2.39	0.96	13.50	0.04	12.10
1981	0.03	0.92	12.64	0.14	13.72	2.94	0.06	1.26	0.04	4.31	−2.92	0.86	11.38	0.10	9.41
1982	0.02	0.95	10.78	0.12	11.86	2.79	0.05	1.73	0.04	4.61	−2.77	0.90	9.05	0.08	7.25
1983	0.03	0.94	10.65	0.13	11.75	2.04	0.06	1.57	0.03	3.69	−2.01	0.89	9.08	0.10	8.06
1984	0.03	0.85	11.43	0.16	12.47	2.15	0.06	1.54	0.03	3.79	−2.12	0.79	9.89	0.12	8.68
1985	0.05	0.95	10.61	0.17	11.78	2.44	0.06	1.66	0.04	4.20	−2.39	0.90	8.95	0.13	7.58
1986	0.06	0.75	13.20	0.15	14.15	2.25	0.06	1.67	0.04	4.02	−2.19	0.69	11.53	0.11	10.13
1987	0.04	0.99	14.16	0.20	15.40	2.09	0.05	1.63	0.03	3.81	−2.05	0.94	12.53	0.17	11.59
1988	0.05	1.30	15.75	0.20	17.30	2.50	0.07	1.74	0.05	4.37	−2.45	1.22	14.01	0.15	12.93
1989	0.07	1.39	17.16	0.15	18.77	2.64	0.11	1.84	0.08	4.66	−2.57	1.28	15.33	0.07	14.11
1990	0.07	1.55	17.12	0.08	18.82	2.77	0.09	1.82	0.07	4.75	−2.70	1.46	15.29	0.01	14.06
1991	0.08	1.80	16.35	0.10	18.33	2.85	0.13	2.13	0.03	5.14	−2.77	1.67	14.22	0.08	13.19
1992	0.10	2.16	16.97	0.15	19.37	2.68	0.22	2.01	0.03	4.94	−2.59	1.94	14.96	0.12	14.44
1993	0.20	2.40	18.51	0.16	21.27	1.96	0.14	2.12	0.04	4.26	−1.76	2.25	16.40	0.12	17.01
1994	0.22	2.68	19.24	0.24	22.39	1.88	0.16	1.99	0.04	4.06	−1.66	2.52	17.26	0.21	18.33
1995	0.24	2.90	18.88	0.24	22.26	2.32	0.16	1.99	0.05	4.51	−2.08	2.74	16.89	0.19	17.75
1996	0.20	3.00	20.29	0.21	23.70	2.37	0.16	2.06	0.05	4.63	−2.17	2.85	18.23	0.16	19.07
1997	0.19	3.06	21.74	0.22	25.22	2.19	0.16	2.10	0.06	4.51	−2.01	2.90	19.64	0.16	20.70
1998	0.22	3.22	22.91	0.23	26.58	2.09	0.16	1.97	0.07	4.30	−1.87	3.06	20.94	0.16	22.28
1999	0.23	3.66	23.13	0.23	27.25	1.53	0.16	1.95	0.07	3.71	−1.30	3.50	21.18	0.16	23.54
2000	0.31	3.87	24.53	0.26	28.97	1.53	0.25	2.15	0.08	4.01	−1.21	3.62	22.38	0.18	24.97
2001	0.49	4.07	25.40	0.19	ᴿ30.16	1.27	0.38	2.04	ᴿ0.09	ᴿ3.77	−0.77	3.69	23.36	ᴿ0.10	26.39
2002	0.42	4.10	ᴿ24.68	0.20	ᴿ29.41	1.03	0.52	2.04	ᴿ0.07	ᴿ3.66	−0.61	3.58	ᴿ22.63	0.14	ᴿ25.74
2003ᴾ	0.63	4.02	26.21	0.17	31.02	1.12	0.70	2.13	0.10	4.05	−0.49	3.32	24.07	0.07	26.97

TABLE 1.2

Energy imports, exports, and net imports, 1949–2003 [CONTINUED]

(Quadrillion Btu)

[1]Includes imports into the Strategic Petroleum reserve, which began in 1977.
[2]Coal coke and small amounts of electricity transmitted across U.S. borders with Canada and Mexico.
R = Revised.
P = Preliminary.
(s) = Less than 0.005 quadrillion Btu and greater than −0.005 quadrillion Btu.
Notes: Includes trade between the United States (50 states and the District of Columbia) and its territories and possessions. Totals or net import items may not equal sum of components due to independent rounding. Web Page: For data not shown for 1951–1969, see http://www.eia.doe.gov/emeu/aer/overview.html.

SOURCE: "Table 1.4. Energy Imports, Exports, and Net Imports, Selected Years, 1949–2003," in *Annual Energy Review 2003*, U.S. Department of Energy, Energy Information Administration, Office of Energy Markets and End Use, September 7, 2004, http://www.eiea.doe.gov/emeu/aer/pdf/aer.pdf (accessed September 28, 2004)

FIGURE 1.8

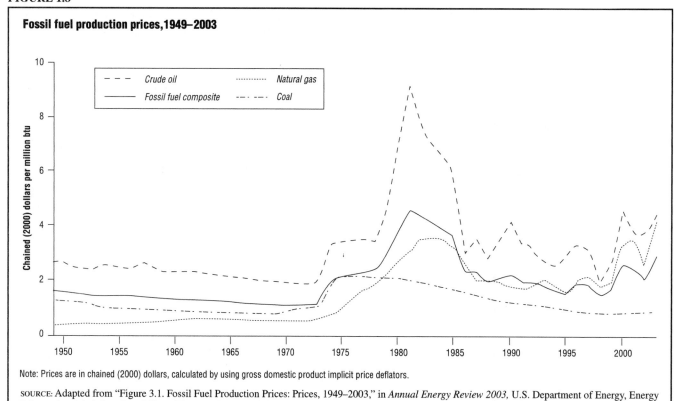

Fossil fuel production prices,1949–2003

Note: Prices are in chained (2000) dollars, calculated by using gross domestic product implicit price deflators.

SOURCE: Adapted from "Figure 3.1. Fossil Fuel Production Prices: Prices, 1949–2003," in *Annual Energy Review 2003,* U.S. Department of Energy, Energy Information Administration, Office of Energy Markets and End Use, September 7, 2004, http://www.eia.doe.gov/emeu/aer/pdf/aer.pdf (accessed September 28, 2004)

FIGURE 1.9

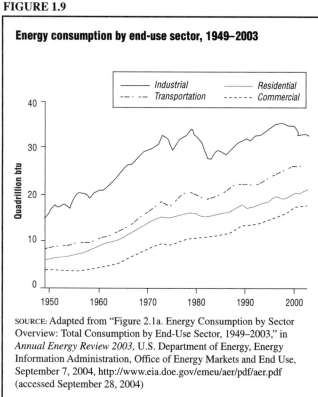

Energy consumption by end-use sector, 1949–2003

SOURCE: Adapted from "Figure 2.1a. Energy Consumption by Sector Overview: Total Consumption by End-Use Sector, 1949–2003," in *Annual Energy Review 2003,* U.S. Department of Energy, Energy Information Administration, Office of Energy Markets and End Use, September 7, 2004, http://www.eia.doe.gov/emeu/aer/pdf/aer.pdf (accessed September 28, 2004)

FIGURE 1.10

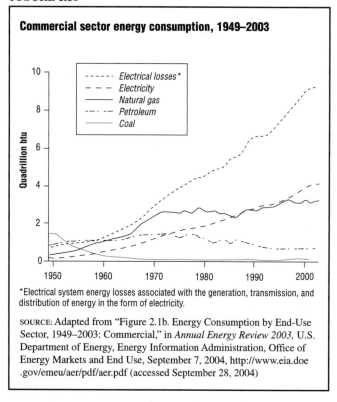

Commercial sector energy consumption, 1949–2003

*Electrical system energy losses associated with the generation, transmission, and distribution of energy in the form of electricity.

SOURCE: Adapted from "Figure 2.1b. Energy Consumption by End-Use Sector, 1949–2003: Commercial," in *Annual Energy Review 2003,* U.S. Department of Energy, Energy Information Administration, Office of Energy Markets and End Use, September 7, 2004, http://www.eia.doe.gov/emeu/aer/pdf/aer.pdf (accessed September 28, 2004)

2003, almost all the energy from the Middle East is in the form of oil or natural gas, while coal is a major source in China. Canada is the leading producer of hydroelectric power and alone accounted for 12% of world production of this form of power in 2002. France produces the highest percentage of its energy from nuclear power.

TABLE 1.3

Fossil fuel production prices, 1949–2003

(Dollars per million btu)

Year	Coal[1]		Natural gas[2]		Crude oil[3]		Fossil fuel composite[4]		
	Nominal	Real[5]	Nominal	Real[5]	Nominal	Real[5]	Nominal[5]	Real[5]	Percent change[6]
1949	0.21	R1.28	0.05	R0.33	0.44	R2.68	0.26	R1.60	—
1950	0.21	R1.25	0.06	R0.38	0.43	R2.62	0.26	R1.54	R−3.7
1955	0.19	R0.99	0.09	R0.48	0.48	R2.55	0.27	R1.45	−3.9
1960	0.19	R0.92	0.13	R0.60	0.50	R2.36	0.28	R1.35	−2.4
1965	0.18	R0.82	0.15	R0.64	0.49	R2.19	0.28	R1.23	R−1.8
1970	0.27	R0.97	0.15	R0.56	0.55	R1.99	0.32	R1.16	R1.0
1971	0.30	R1.05	0.16	R0.56	0.58	R2.02	0.34	R1.18	1.8
1972	0.33	R1.09	0.17	R0.57	0.58	R1.94	0.35	R1.16	R−1.4
1973	0.37	R1.15	0.20	R0.63	0.67	R2.11	0.40	R1.25	R7.8
1974	0.69	R1.98	0.27	R0.79	1.18	R3.41	0.68	R1.95	55.8
1975	0.85	R2.22	0.40	R1.06	1.32	R3.48	0.82	R2.16	R10.9
1976	0.86	R2.13	0.53	R1.32	1.41	R3.51	0.90	R2.24	3.9
1977	0.88	R2.07	0.72	R1.69	1.48	R3.46	1.01	R2.36	R5.1
1978	0.98	R2.15	0.84	R1.83	1.55	R3.39	1.12	R2.44	R3.3
1979	1.06	R2.14	1.08	R2.18	2.18	R4.40	1.42	R2.86	R17.4
1980	1.10	R2.04	1.45	R2.68	3.72	R6.89	2.04	R3.78	R32.1
1981	1.18	R2.00	1.80	R3.04	5.48	R9.27	2.75	R4.64	22.9
1982	1.23	R1.95	2.22	R3.54	4.92	R7.84	2.76	R4.40	R−5.3
1983	1.18	R1.81	2.32	R3.56	4.52	R6.93	2.70	R4.14	−5.8
1984	1.16	R1.72	2.40	R3.55	4.46	R6.60	2.65	R3.91	R−5.6
1985	1.15	R1.65	2.26	R3.24	4.15	R5.96	2.51	R3.60	R−7.9
1986	1.09	R1.52	1.75	R2.45	2.16	R3.03	1.65	R2.32	−35.6
1987	1.05	R1.44	1.50	R2.05	2.66	R3.63	1.70	R2.32	R0.1
1988	1.01	R1.34	1.52	R2.01	2.17	R2.87	1.53	R2.03	−12.8
1989	1.00	R1.28	1.53	R1.94	2.73	R3.48	1.67	R2.13	5.0
1990	1.00	R1.22	1.55	R1.90	3.45	R4.23	1.84	R2.26	R6.2
1991	0.99	R1.17	1.48	R1.75	2.85	R3.38	1.67	R1.98	R−12.4
1992	0.97	R1.12	1.57	R1.82	2.76	R3.19	1.66	R1.92	R−3.0
1993	0.93	R1.05	1.84	R2.09	2.46	R2.78	1.67	R1.89	R−1.5
1994	0.91	R1.01	1.67	R1.86	2.27	R2.52	1.53	R1.69	−10.5
1995	0.88	R0.96	1.40	R1.52	2.52	R2.74	1.47	R1.60	−5.5
1996	0.87	R0.93	1.96	R2.09	3.18	R3.39	1.82	R1.94	R21.4
1997	0.85	R0.89	2.10	R2.20	2.97	R3.11	1.81	R1.90	R−2.4
1998	0.83	R0.86	1.77	R1.83	1.87	R1.94	1.41	R1.46	R−22.8
1999	0.79	R0.81	1.98	R2.02	2.68	R2.74	1.65	R1.69	15.4
2000	0.80	R0.80	R3.32	R3.32	4.61	R4.61	R2.60	R2.60	R54.2
2001	R0.83	R0.81	R3.62	R3.54	3.77	R3.68	R2.53	R2.47	R−5.1
2002	R0.87	R0.83	2.67	R2.56	3.88	R3.73	R2.21	R2.12	R−14.0
2003P	0.86	0.82	4.50	4.26	4.75	4.50	3.12	2.95	39.1

[1]Free-on-board (f.o.b.) rail/barge prices, which are the f.o.b. prices of coal at the point of first sale, excluding freight or shipping and insurance costs.
[2]Wellhead prices.
[3]Domestic first purchase prices.
[4]Derived by multiplying the price per btu of each fossil fuel by the total btu content of the production of each fossil fuel and dividing this accumulated value of total fossil fuel production by the accumulated btu content of total fossil fuel production.
[5]In chained (2000) dollars, calculated by using gross domestic product implicit price deflators.
[6]Based on real values.
R=Revised.
P=Preliminary.
—=Not applicable.
Web page: For data not shown for 1951–1969, see http://www.eia.doe.gov/emeu/aer/finan.html.

SOURCE: "Table 3.1. Fossil Fuel Production Prices, Selected Years, 1949–2003 (Dollars per Million Btu)," in *Annual Energy Review 2003,* U.S. Department of Energy, Energy Information Administration, Office of Energy Markets and End Use, September 7, 2004, http://www.eia.doe.gov/emeu/aer/pdf/aer.pdf (accessed September 28, 2004)

World Consumption

Table 1.5 shows the world consumption of energy by region (including the United States, China, Russia, Japan, and Germany) from 1981 to 2002. The five countries singled out on the table together consumed 50% of the world's total energy supply in 2002. The United States, by far the world's largest consumer of energy, used about 97.6 quadrillion Btu in 2002, or nearly one-fourth of the energy consumed worldwide. This amount was more than twice China's 43.2 quadrillion Btu, while Russia consumed 27.5 quadrillion Btu.

FUTURE TRENDS IN ENERGY CONSUMPTION, PRODUCTION, AND PRICES

The EIA forecasts energy supply, demand, and prices every year in its *Annual Energy Outlook,* which is used by decision makers in the public and private sectors. The

FIGURE 1.11

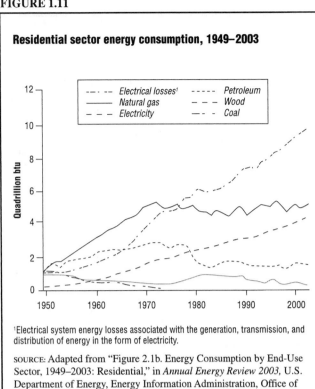

Residential sector energy consumption, 1949–2003

¹Electrical system energy losses associated with the generation, transmission, and distribution of energy in the form of electricity.

SOURCE: Adapted from "Figure 2.1b. Energy Consumption by End-Use Sector, 1949–2003: Residential," in *Annual Energy Review 2003,* U.S. Department of Energy, Energy Information Administration, Office of Energy Markets and End Use, September 7, 2004, http://www.eia.doe.gov/emeu/aer/pdf/aer.pdf (accessed September 28, 2004)

FIGURE 1.12

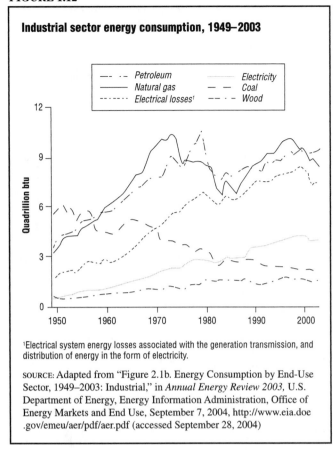

Industrial sector energy consumption, 1949–2003

¹Electrical system energy losses associated with the generation transmission, and distribution of energy in the form of electricity.

SOURCE: Adapted from "Figure 2.1b. Energy Consumption by End-Use Sector, 1949–2003: Industrial," in *Annual Energy Review 2003,* U.S. Department of Energy, Energy Information Administration, Office of Energy Markets and End Use, September 7, 2004, http://www.eia.doe.gov/emeu/aer/pdf/aer.pdf (accessed September 28, 2004)

EIA's latest projections, through 2025, are based on current U.S. laws, regulations, and economic conditions.

Total energy consumption in the United States is projected to increase from 97.7 to 136.5 quadrillion Btu between 2002 and 2025, an average annual increase of 1.5%. (See Table 1.6.) That projection becomes higher with high economic growth and/or low world oil prices, and lower with low economic growth and/or high world oil prices. In addition, energy consumption will increase in all end-use sectors to 2025.

The consumption of petroleum, natural gas, coal, and nonhydroelectric renewable energy sources is expected to rise significantly from 2002 to 2025 (see Figure 1.16), collectively increasing at an average annual rate of 1.8%. Total petroleum demand is projected to increase from 19.6 million barrels of oil per day in 2002 to 28.3 million barrels per day in 2025. Coal, natural gas, and renewable fuels consumption is projected to grow in part to meet the increased demand for electricity. Consumption of hydroelectric power and electricity generated from nuclear power will remain steady.

The rising consumption of petroleum by Americans is projected to lead to increasing petroleum imports by the United States through 2025. (See Figure 1.17.) Gross oil imports are projected to increase from 11.5 million barrels per day in 2002 to 20.7 million barrels per day in 2025. Most of the increase in imports will be crude oil because distillation capacity in the United States is expected to increase. Nevertheless, net imports of refined petroleum products are still expected to more than double over the next two decades.

In the period 2002–25 electricity prices in the United States are projected to decline slightly and then level out because of restructuring laws designed to increase competition in the industry, though restructuring has slowed. As shown in Figure 1.18, the electricity price projections in *Annual Energy Outlook 2004* (AEO2004) are only slightly higher than the projections from *Annual Energy Outlook 2003* (AEO2003). Figure 1.18 also shows that coal prices are expected to remain steady from 2002 to 2025, while crude oil and natural gas prices are expected to decrease from 2002 levels, then slowly rise through 2025.

FIGURE 1.13

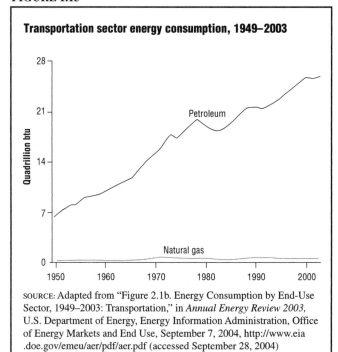

Transportation sector energy consumption, 1949–2003

SOURCE: Adapted from "Figure 2.1b. Energy Consumption by End-Use Sector, 1949–2003: Transportation," in *Annual Energy Review 2003,* U.S. Department of Energy, Energy Information Administration, Office of Energy Markets and End Use, September 7, 2004, http://www.eia .doe.gov/emeu/aer/pdf/aer.pdf (accessed September 28, 2004)

FIGURE 1.14

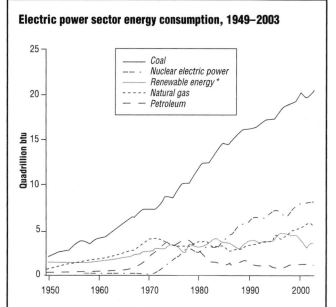

Electric power sector energy consumption, 1949–2003

*Conventional hydroelectric power, wood, waste, geothemal, solar, and wind.

SOURCE: Adapted from "Figure 2.1a. Energy Consumption by Sector Overview: Electric Power Sector, 1949–2003," in *Annual Energy Review 2003,* U.S. Department of Energy, Energy Information Administration, Office of Energy Markets and End Use, September 7, 2004, http://www.eia.doe.gov/emeu/aer/pdf/aer.pdf (accessed September 28, 2004)

FIGURE 1.15

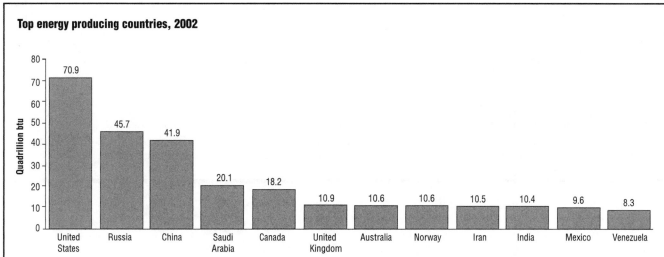

Top energy producing countries, 2002

SOURCE: Adapted from "Figure 11.2. World Primary Energy Production by Region and Country: Top Producing Countries, 2002," in *Annual Energy Review 2003,* U.S. Department of Energy, Energy Information Administration, Office of Energy Markets and End Use, September 7, 2004, http://www.eia.doe.gov/ emeu/aer/pdf/aer.pdf (accessed September 28, 2004)

TABLE 1.4

World primary energy production by source, 1970–2002

(Quadrillion btu)

Year	Coal	Natural gas[1]	Crude oil[2]	Natural gas plant liquids	Nuclear electric power[3]	Hydroelectric power[3]	Geothermal[3] and other[4]	Total
1970	62.96	37.09	97.09	3.61	0.90	12.15	1.59	215.39
1971	61.72	39.80	102.70	3.85	1.23	12.74	1.61	223.64
1972	63.65	42.08	108.52	4.09	1.66	13.31	1.68	234.99
1973	63.87	44.44	117.88	4.23	2.15	13.52	1.73	247.83
1974	63.79	45.35	117.82	4.22	2.86	14.84	1.76	250.64
1975	66.20	45.67	113.08	4.12	3.85	15.03	1.74	249.69
1976	67.32	47.62	122.92	4.24	4.52	15.08	1.97	263.67
1977	68.46	48.85	127.75	4.40	5.41	15.56	2.11	272.54
1978	69.56	50.26	128.51	4.55	6.42	16.80	2.32	278.41
1979	73.83	53.93	133.87	4.87	6.69	17.69	2.48	293.36
1980	R72.54	54.73	128.12	5.10	7.58	R18.04	2.95	R289.05
1981	R72.91	55.56	120.16	5.36	8.53	R18.41	R3.10	R284.02
1982	R75.55	55.49	114.51	5.34	9.51	R18.88	3.24	R282.53
1983	R75.58	56.12	113.97	5.34	10.72	R19.88	3.51	R285.13
1984	R79.73	61.78	116.86	5.71	R12.99	R20.38	3.64	R301.10
1985	R83.54	64.22	115.40	5.82	15.30	R20.62	3.67	R308.56
1986	R85.62	65.32	120.24	6.12	16.25	R21.08	R3.74	R318.37
1987	R87.41	68.48	121.16	6.32	17.64	R21.11	3.79	R325.92
1988	R89.25	71.80	125.93	6.63	19.23	R21.72	R3.93	R338.50
1989	R90.67	74.24	127.98	6.67	19.74	R21.77	4.29	R345.37
1990	R92.04	75.87	129.50	6.85	20.31	R22.54	R3.96	R351.08
1991	R87.32	76.69	128.77	7.13	21.13	R23.04	4.04	R348.13
1992	R86.74	76.90	129.13	7.38	21.23	R22.80	R4.32	R348.50
1993	R84.08	78.41	128.86	7.68	21.96	R24.10	R4.35	R349.43
1994	R86.14	79.18	130.46	7.85	22.36	R24.21	R4.55	R354.75
1995	R88.71	80.24	133.32	8.16	23.21	R25.43	R4.76	R363.84
1996	R88.55	83.94	136.64	8.31	24.05	R25.96	R4.88	R372.33
1997	R92.41	83.89	140.52	8.51	23.82	R26.18	R4.92	R380.26
1998	R91.08	85.58	143.15	8.75	24.34	R26.22	R4.83	R383.94
1999	R90.61	R87.53	140.79	R9.01	25.08	R26.68	R5.07	R384.77
2000	R91.44	R91.03	146.50	R9.43	R25.52	R27.12	R5.24	R396.28
2001	R97.13	R93.38	R145.25	R10.07	R26.40	R26.02	R5.09	R403.33
2002[P]	97.56	95.20	142.86	10.55	26.85	26.59	5.52	405.12

[1]Dry production.
[2]Includes lease condensate.
[3]Net generation, i.e., gross generation less plant use.
[4]Includes net electricity generation from wood, waste, solar, and wind. Data for the United States also include other renewable energy.
R=Revised.
P=Preliminary.
Totals may not equal sum of components due to independent rounding.
Web Page: For related information, see http://www.eia.doe.gov/international.
Sources: 1971–1979—Energy Information Administration (EIA), International Energy Database. 1980 forward—EIA, "International Energy Annual 2002" (May 2004)

SOURCE: "Table 11.1. World Primary Energy Production by Source, 1970–2002 (Quadrillion Btu)," in *Annual Energy Review 2003*, U.S. Department of Energy, Energy Information Administration, Office of Energy Markets and End Use, September 7, 2004, http://www.eia.doe.gov/emeu/aer/pdf/aer.pdf (accessed September 28, 2004)

TABLE 1.5

World primary energy consumption, 1981–2002

(Quadrillion btu)

Region/country	1981	1984	1987	1990	1993	1996	1999	2002
United States	76.368	76.726	79.189	84.606	87.578	94.225	96.774	97.649
North America	**90.086**	**90.887**	**94.485**	**100.906**	**104.734**	**112.365**	**115.465**	**117.356**
Antarctica	0.002	0.003	0.003	0.003	0.003	0.003	0.003	0.003
Antigua and Barbuda	0.005	0.005	0.006	0.006	0.006	0.007	0.007	0.007
Central & South America	**11.500**	**12.108**	**13.462**	**14.527**	**16.132**	**18.527**	**20.305**	**21.186**
Germany	NA	NA	NA	NA	14.065	14.353	14.117	14.269
Germany, East	3.560	3.518	3.827	3.358	NA	NA	NA	NA
Germany, West	10.769	10.914	11.405	11.460	NA	NA	NA	NA
Western Europe	**56.716**	**57.685**	**62.021**	**64.053**	**64.654**	**68.201**	**70.320**	**72.265**
Russia	NA	NA	NA	NA	31.964	27.374	26.772	27.536
Eastern Europe & Former U.S.S.R.	**61.226**	**67.952**	**73.927**	**74.116**	**60.012**	**51.809**	**50.053**	**51.904**
Middle East	**6.183**	**7.880**	**9.704**	**11.084**	**12.733**	**14.610**	**16.601**	**18.867**
Africa	**6.996**	**8.327**	**8.834**	**9.296**	**9.953**	**10.921**	**11.600**	**12.750**
China	17.192	20.453	24.755	27.001	31.317	36.081	36.995	43.177
Japan	15.127	15.651	16.201	18.273	19.397	21.252	21.657	21.965
Asia & Oceania	**49.733**	**56.388**	**64.646**	**74.227**	**85.062**	**98.387**	**104.417**	**116.868**
World total	**282.440**	**301.227**	**327.080**	**348.208**	**353.281**	**374.820**	**388.761**	**411.196**

Notes: Data for the most recent year are preliminary. Total primary energy consumption reported in this table includes the consumption of petroleum, dry natural gas, coal, and net hydroelectric, nuclear, and geothermal, solar, wind, and wood and waste electric power. Total primary energy consumption for each country also includes net electricty imports (electricity imports minus electricity exports).

SOURCE: Adapted from "Table E1. World Total Primary Energy Consumption (Quadrillion Btu), 1980–2002," in *International Energy Annual 2002,* U.S. Department of Energy, Energy Information Administration, May 28, 2004, http://www.eia.doe.gov/pub/international/iealf/tablee1.xls (accessed November 8, 2004)

FIGURE 1.16

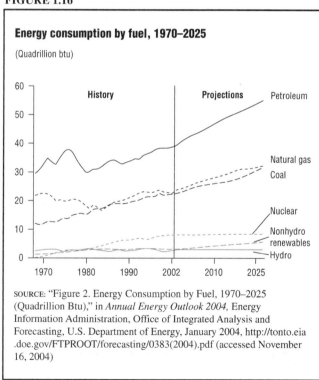

Energy consumption by fuel, 1970–2025

(Quadrillion btu)

SOURCE: "Figure 2. Energy Consumption by Fuel, 1970–2025 (Quadrillion Btu)," in *Annual Energy Outlook 2004,* Energy Information Administration, Office of Integrated Analysis and Forecasting, U.S. Department of Energy, January 2004, http://tonto.eia .doe.gov/FTPROOT/forecasting/0383(2004).pdf (accessed November 16, 2004)

FIGURE 1.17

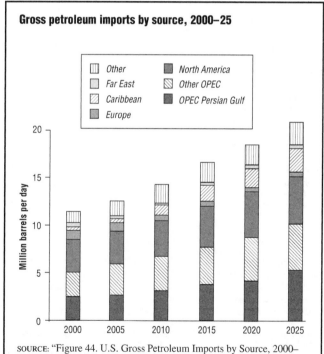

Gross petroleum imports by source, 2000–25

SOURCE: "Figure 44. U.S. Gross Petroleum Imports by Source, 2000– 2025 (Million Barrels per Day)," in *Annual Energy Outlook 2004,* U.S. Department of Energy, Energy Information Administration, Office of Integrated Analysis and Forecasting, January 2004, http://tonto.eia.doe .gov/FTPROOT/forecasting/0383(2004).pdf (accessed November 16, 2004)

TABLE 1.6

Summary of projected total energy supply and disposition, 2001–25

Energy and economic factors	2001	2002	2010	2015	2020	2025	Average annual change, 2002–2025
Primary energy production (quadrillion btu)							
Petroleum	14.70	14.47	15.66	14.91	13.95	13.24	−0.4%
Dry natural gas	20.23	19.56	21.05	22.20	24.43	24.64	1.0%
Coal	23.97	22.70	25.25	26.14	27.92	31.10	1.4%
Nuclear power	8.03	8.15	8.29	8.48	8.53	8.53	0.2%
Renewable energy	5.25	5.84	7.18	7.84	8.45	9.00	1.9%
Other	0.53	1.13	0.88	0.79	0.81	0.84	−1.3%
Total	**72.72**	**71.85**	**78.30**	**80.36**	**84.09**	**87.33**	**0.9%**
Net imports (quadrillion btu)							
Petroleum	23.29	22.56	28.13	33.20	37.25	41.69	2.7%
Natural gas	3.69	3.58	5.63	6.39	6.63	7.41	3.2%
Coal/other (−indicates export)	−0.67	−0.51	0.06	0.26	0.43	0.61	NA
Total	**26.31**	**25.63**	**33.82**	**39.84**	**44.31**	**49.71**	**2.9%**
Consumption (quadrillion btu)							
Petroleum products	38.49	38.11	44.15	48.26	51.35	54.99	1.6%
Natural gas	23.05	23.37	26.82	28.74	31.21	32.21	1.4%
Coal	22.04	22.18	25.23	26.32	28.30	31.73	1.6%
Nuclear power	8.03	8.15	8.29	8.48	8.53	8.53	0.2%
Renewable energy	5.25	5.84	7.18	7.84	8.46	9.00	1.9%
Other	0.08	0.07	0.11	0.11	0.07	0.03	−4.6%
Total	**96.94**	**97.72**	**111.77**	**119.75**	**127.92**	**136.48**	**1.5%**
Petroleum (million barrels per day)							
Domestic crude production	5.74	5.62	5.93	5.53	4.95	4.61	−0.9%
Other domestic production	3.11	3.60	3.59	3.72	3.98		0.4%
Net imports	10.90	10.54	13.17	15.52	17.48	19.67	2.7%
Consumption	19.71	19.61	22.71	24.80	26.41	28.30	1.6%
Natural gas (trillion cubic feet)							
Production	19.79	19.13	20.59	21.72	23.89	24.08	1.0%
Net imports	3.60	3.49	5.50	6.24	6.47	7.24	3.2%
Consumption	22.48	22.78	26.15	28.03	30.44	31.41	1.4%
Coal (million short tons)							
Production	1,138	1,105	1,230	1,285	1,377	1,543	1.5%
Net imports	−29	−23	−2	6	14	23	NA
Consumption	1,060	1,066	1,229	1,291	1,391	1,567	1.7%
Prices (2002 dollars)							
World oil price (dollars per barrel)	22.25	23.68	24.17	25.07	26.02	27.00	0.6%
Domestic natural gas at wellhead (dollars per thousand cubic feet)	4.14	2.95	3.40	4.19	4.28	4.40	1.8%
Domestic coal at minemouth (dollars per short ton)	17.79	17.90	16.88q	16.4	16.32	16.57	−0.3%
Average electricity price (cents per kilowatthour)	7.4	7.2	6.6	6.8	6.9	6.9	−0.2%
Economic indicators							
Real gross domestic product (billion 1996 dollars)	9,215	9,440	12,190	14,101	16,188	18,520	3.0%
GDP chain-type price index (index, 1996=1.000)	1.094	1.107	1.301	1.503	1.774	2.121	2.9%
Real disposable personal income (billion 1996 dollars)	6,748	7,032	8,894	10,33	11,86	13,826	3.0%
Value of manufacturing shipments (billion 1996 dollars)	5,368	5,285	6,439	7,345	8,344	9,491	2.6%
Energy intensity (thousand btu per 1996 dollar of GDP)	**10.53**	**10.36**	**9.17**	**8.50**	**7.91**	**7.37**	**−1.5%**
Carbon dioxide emissions (million metric tons)	**5,691.7**	**5,729.3**	**6,558.8**	**7,028.4**	**7,535.6**	**8,142.0**	**1.5%**

Notes: Quantities are derived from historical volumes and assumed thermal conversion factors. Other production includes liquid hydrogen, methanol, supplemental natural gas, and some inputs to refineries. Net imports of petroleum include crude oil, petroleum products, unfinished oils, alcohols, ethers, and blending components. Other net imports include coal coke and electricity. Some refinery inputs appear as petroleum product consumption. Other consumption includes net electricity imports, liquid hydrogen, and methanol.

SOURCE: "Table 1. Total Energy Supply and Disposition in the AEO2004 Reference Case: Summary, 2001–2025," in *Annual Energy Outlook 2004*, U.S. Department of Energy, Energy Information Administration, Office of Integrated Analysis and Forecasting, January 2004, http://tonto.eia.doe.gov/FTPROOT/forecasting/0383(2004).pdf (accessed November 16, 2004)

FIGURE 1.18

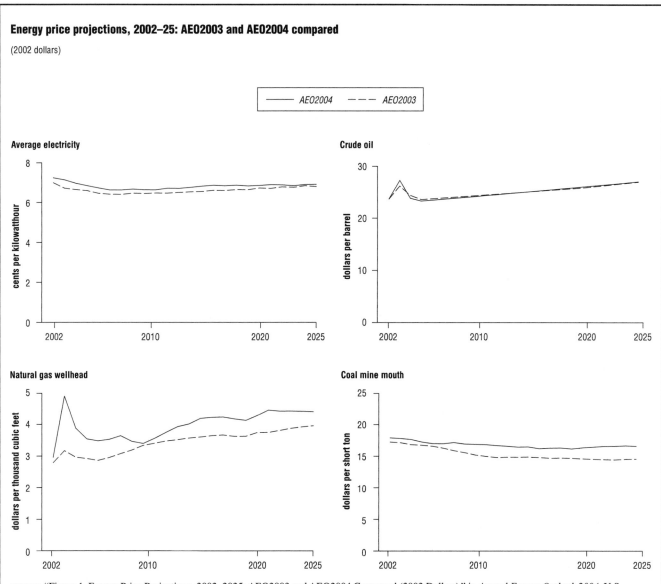

Energy price projections, 2002–25: AEO2003 and AEO2004 compared

(2002 dollars)

——— AEO2004　　- - - - AEO2003

Average electricity

Crude oil

Natural gas wellhead

Coal mine mouth

SOURCE: "Figure 1. Energy Price Projections, 2002–2025: AEO2003 and AEO2004 Compared (2002 Dollars)," in *Annual Energy Outlook 2004,* U.S. Department of Energy, Energy Information Administration, Office of Integrated Analysis and Forecasting, January 2004, http://tonto.eia.doe.gov/ FTPROOT/forecasting/0383(2004).pdf (accessed November 16, 2004)

CHAPTER 2

OIL

THE QUEST FOR OIL

On August 27, 1859, Edwin Drake struck oil sixty-nine feet below the surface of the earth near Titusville, Pennsylvania. This was the first successful modern oil well, which ushered in the "Age of Petroleum." Not only did petroleum help meet the growing demand for new and better fuels for heating and lighting, but it also proved to be an excellent fuel for the internal combustion engine, which was developed in the late 1800s.

Sources of Oil

Almost all oil comes from underground reservoirs. The most widely accepted explanation of how oil and gas are formed within the earth is that these fuels are the products of intense heat and pressure applied over millions of years to organic (formerly alive) sediments buried in geological formations. For this reason they are called fossil fuels. They are limited (nonrenewable) resources, which means that they are formed much more slowly than they are used, so they are finite in supply.

At one time it was believed that crude oil flowed in underground streams and accumulated in lakes or caverns in the earth. Today, scientists know that a petroleum reservoir is usually a solid sandstone or limestone formation overlaid with a layer of impermeable rock or shale, which creates a shield. The petroleum accumulates within the pores and fractures of the rock and is trapped beneath the seal. Anticlines (archlike folds in a bed of rock), faults, and salt domes are common trapping formations. (See Figure 2.1.) Oil and natural gas deposits can be found at varying depths. Wells are drilled to reach the reservoirs and extract the oil. Deep wells are more expensive to drill and are usually attempted only to reach large reservoirs or when the price of oil is high.

How Oil Is Drilled and Recovered

Oil is crucial to the economies of industrialized nations, and oil production is strongly influenced by market prices.

FIGURE 2.1

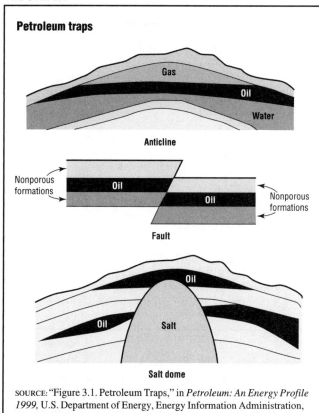

Petroleum traps

Anticline

Fault

Salt dome

SOURCE: "Figure 3.1. Petroleum Traps," in *Petroleum: An Energy Profile 1999,* U.S. Department of Energy, Energy Information Administration, July 1999, http://www.eia.doe.gov/pub/oil_gas/petroleum/analysis_publications/petroleum_profile_1999/profile99v8.pdf (accessed November 8, 2004)

Scouting for new deposits and drilling exploratory wells is expensive. It is also costly to drill, maintain, and operate production wells. If the price of oil falls too low, operators will shut off expensive wells because they cannot recover their higher operating costs. In general, when oil prices are high, oil companies drill; when prices are low, drilling drops off.

Figure 2.2 shows a diagram of a rotary drilling system, or rotary rig. The rotating bit at the end of lengths

FIGURE 2.2

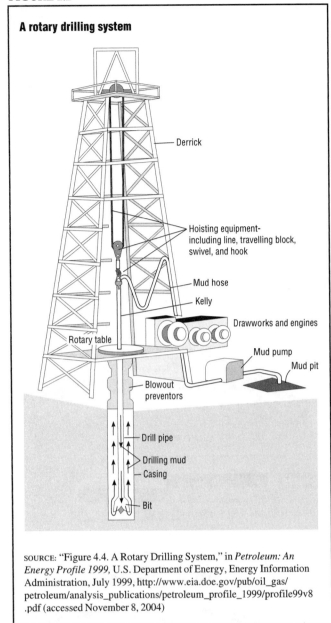

A rotary drilling system

Derrick

Hoisting equipment-
including line, travelling block,
swivel, and hook

Mud hose

Kelly

Drawworks and engines

Rotary table

Mud pump

Mud pit

Blowout
preventors

Drill pipe

Drilling mud

Casing

Bit

SOURCE: "Figure 4.4. A Rotary Drilling System," in *Petroleum: An Energy Profile 1999*, U.S. Department of Energy, Energy Information Administration, July 1999, http://www.eia.doe.gov/pub/oil_gas/ petroleum/analysis_publications/petroleum_profile_1999/profile99v8 .pdf (accessed November 8, 2004)

of pipe drills a hole into the ground. Drilling mud is pushed through the pipe and the drill bit, which forces the mud and bits of rock from the drilling process back to the surface, as shown by the "up" arrows in the diagram. As the well gets deeper, more pipe is added. The oil derrick supports equipment that can lift the pipe and drill bit from the well when drill bits need to be changed or replaced.

Oil deposits are not distributed evenly over the world, and some that have been tapped for decades are being exhausted. As an oil reservoir is depleted, various techniques can be used to recover additional petroleum. These include the injection of water, chemicals, or steam to force more oil from the rock. These recovery techniques can be expensive and are used only when the price of oil is relatively high.

TYPES OF OIL

While crude oil is usually dark when it comes from the ground, it may also be green, red, yellow, or colorless, depending on its chemical composition and the amount of sulfur, oxygen, nitrogen, and trace minerals present. Its viscosity (thickness, or resistance to flow) can range from as thin as water to as thick as tar. Crude oil is refined, or chemically processed, into finished petroleum products; it has limited uses in its natural form.

Crude oils vary in quality. "Sweet" crudes have little sulfur, refine easily, and are worth more than "sour" crudes, which contain more impurities. "Light" crudes, which have more short molecules, yield more gasoline and are more profitable than "heavy" crudes, which have more long molecules and bring a lower price in the market.

In addition to crude oil, there are two other sources of petroleum: lease condensate and natural gas plant liquids. Lease condensate is a liquid recovered from natural gas. It consists primarily of chemical compounds called penthanes and heavier hydrocarbons and is generally blended with crude oil for refining. Natural gas plant liquids are natural gas liquids that are recovered during the refinement of natural gas in processing plants.

USES FOR OIL

Many of the uses for petroleum are well known: gasoline, diesel fuel, jet fuel, and lubricants for transportation; heating oil, residual oil, and kerosene for heat; and heavy residuals for paving and roofing. Petroleum by-products are also vital to the chemical industry, ending up in many different foams, plastics, synthetic fabrics, paints, dyes, inks, and even pharmaceutical drugs. Many chemical plants, because of their dependence on petroleum, are directly connected by pipelines to nearby refineries.

HOW OIL IS REFINED

Before oil can be used by consumers, crude oil, lease condensate, and natural gas plant liquids must be processed into finished products. The first step in refining is distillation, in which crude oil molecules are separated according to size and weight.

During distillation, crude oil is heated until it turns to vapor. (See Figure 2.3.) The vaporized crude enters the bottom of a distillation column, where it rises and condenses on trays. The lightest vapors, such as those of gasoline, rise to the top. The middleweight vapors, such as those of kerosene, rise about halfway up the column. The heaviest vapors, such as those of heavy gas oil, stay at the bottom. The vapors at each level are then condensed to liquid as they are cooled. These liquids are drawn off, and the processes of cracking and reforming further refine each portion. Cracking converts the heaviest fractions of separated petroleum into lighter fractions to produce jet fuel,

TABLE 2.1

Refinery capacity and utilization, selected years, 1949–2003

Year	Operable refineries		Gross input to distillation units (thousand barrels per day)	Utilization[3] (percent)
	Number[1]	Capacity[2] (thousand barrels per day)		
1949	336	6,231	5,556	89.2
1950	320	6,223	5,980	92.5
1955	296	8,386	7,820	92.2
1960	309	9,843	8,439	85.1
1965	293	10,420	9,557	91.8
1970	276	12,021	11,517	92.6
1971	272	12,860	11,881	90.9
1972	274	13,292	12,431	92.3
1973	268	13,642	13,151	93.9
1974	273	14,362	12,689	86.6
1975	279	41,961	12,902	85.5
1976	276	15,237	13,884	87.8
1977	282	16,398	14,982	89.6
1978	296	17,048	15,071	87.4
1979	308	17,441	14,955	84.4
1980	319	17,988	13,796	75.4
1981	324	18,621	12,752	68.6
1982	301	17,890	12,172	69.9
1983	258	16,859	11,947	71.7
1984	247	16,137	12,216	76.2
1985	223	15,659	12,165	77.6
1986	216	15,459	12,826	82.9
1987	219	15,566	13,003	83.1
1988	213	15,915	13,447	84.7
1989	204	15,655	13,551	86.6
1990	205	15,572	13,610	87.1
1991	202	15,676	13,508	86.0
1992	199	15,696	13,600	87.9
1993	187	15,121	13,851	91.5
1994	179	15,034	14,032	92.6
1995	175	15,434	14,119	92.0
1996	170	15,333	14,337	94.1
1997	164	15,452	14,838	95.2
1998	163	15,711	15,113	96.5
1999	159	16,261	15,080	92.6
2000	158	16,512	15,299	92.6
2001	155	16,595	15,369	92.6
2002	153	16,785	R15,180	R90.7
2003P	149	16,757	15,505	92.5

[1]Through 1956, includes only those refineries in operation on January 1; beginning in 1957, includes all "operable" refineries on January 1.
[2]Capacity on January 1.
[3]Through 1980, utilization is derived by dividing gross input to distillation units by one-half of the current year Januray 1 capacity and the following year January 1 capacity. Percentages were derived from unrounded numbers. Beginning in 1981, utilization is derived by averaging reported monthly utilization.
R=Revised.
P=Preliminary.
Web Pages: For data not shown for 1951–1969, see http://www.eia.doe./emeu/aer/petro.html. For related information, see http://www.eia.doe.gov/oil_gas/petroleum/info_glance/petroleum.html.

SOURCE: "Table 5.9. Refinery Capacity and Utilization, Selected Years, 1949–2003," in *Annual Energy Review 2003*, U.S. Department of Energy, Energy Information Administration, Office of Energy Markets and End Use, September 7, 2004, http://www.eia.doe.gov/emeu/aer/pdf/aer.pdf (accessed September 28, 2004)

motor gasoline, home heating oil, and less-residual fuel oils, which are heavier and used for naval ships, commercial and industrial heating, and some power generation. Reforming is used to increase the octane rating of gasoline.

Refining is a continuous process, with crude oil entering the refinery at the same time that finished products leave by pipeline, truck, and train. Although storage tanks surround refineries, they have limited storage capacity. If there is a malfunction and products cannot be processed, they may be burned off (flared) if no storage facility is available. While a small flare is normal at a refinery or a chemical plant, a large flare, or many flares, likely indicates a processing problem.

REFINERY NUMBERS AND CAPACITY

In 2003, 149 refineries were operating in the United States, a drop from 336 in 1949 and 324 in 1981. (See Table 2.1.) Refinery capacity in 2003 was about 16.8 million barrels per day, below the 1981 peak of 18.6 million barrels. As of 2003 U.S. refineries were operating near full capacity. Utilization rates generally increased from a low of 68.6% in 1981—a period of low demand because

FIGURE 2.3

Crude oil distillation

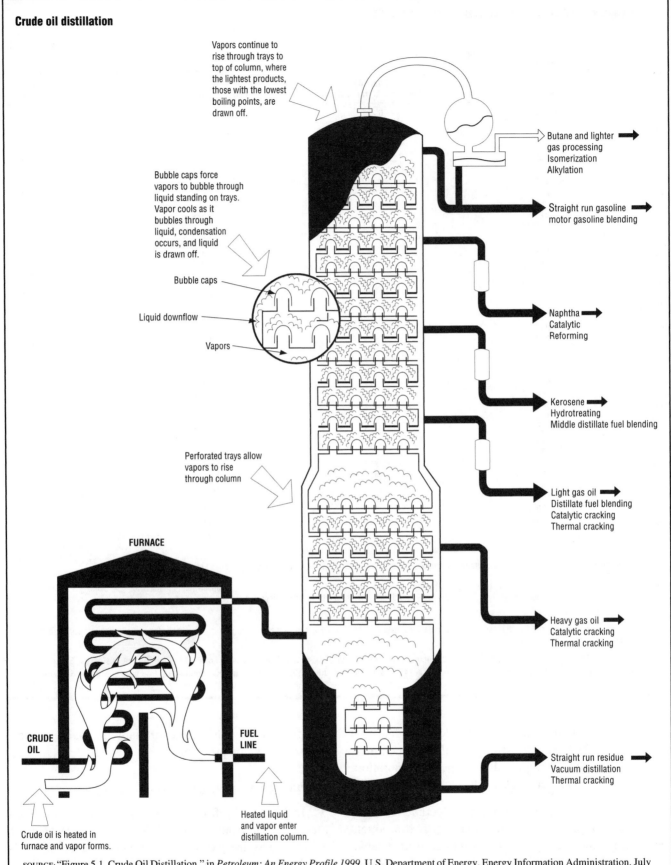

Vapors continue to rise through trays to top of column, where the lightest products, those with the lowest boiling points, are drawn off.

Bubble caps force vapors to bubble through liquid standing on trays. Vapor cools as it bubbles through liquid, condensation occurs, and liquid is drawn off.

Bubble caps

Liquid downflow

Vapors

Perforated trays allow vapors to rise through column

FURNACE

CRUDE OIL

FUEL LINE

Crude oil is heated in furnace and vapor forms.

Heated liquid and vapor enter distillation column.

Butane and lighter gas processing
Isomerization
Alkylation

Straight run gasoline
motor gasoline blending

Naphtha
Catalytic
Reforming

Kerosene
Hydrotreating
Middle distillate fuel blending

Light gas oil
Distillate fuel blending
Catalytic cracking
Thermal cracking

Heavy gas oil
Catalytic cracking
Thermal cracking

Straight run residue
Vacuum distillation
Thermal cracking

SOURCE: "Figure 5.1. Crude Oil Distillation," in *Petroleum: An Energy Profile 1999,* U.S. Department of Energy, Energy Information Administration, July 1999, http://www.eia.doe.gov/pub/oil_gas/petroleum/analysis_publications/petroleum_profile_1999/profile99v8.pdf (accessed November 8, 2004)

of economic recession—to a high of 95.6% in 1998. Although capacity fell slightly in the years since 1998, it was still high in 2003.

As Table 2.1 shows, fewer refineries were operating in the United States in the early twenty-first century than in the past, but they were working at near maximum levels. One reason for the drop in the number of U.S. refineries is that the petroleum industry began shutting down older, inefficient refineries and concentrating production in more efficient plants, which tended to be newer and larger.

Consolidation within the industry has also played a role in refinery operation. For example, the merger of Gulf Oil Corporation into Chevron Corporation in 1984 led to the closing of two large refineries, one in Bakersfield, California, and the other in Cincinnati, Ohio. In 1998 Exxon merged with Mobil Oil and BP merged with Amoco. BP Amoco then bought Arco in April 2000 to create the world's largest non-OPEC (Organization of the Petroleum Exporting Countries) oil producer and the third-largest natural gas producer. In the May 2004 article "Effects of Mergers and Market Concentration in the U.S. Petroleum Industry," in *GAO Highlights*, the General Accounting Office noted that "over 2,600 mergers have occurred in the U.S. petroleum industry since the 1990s." The report said that the majority of the mergers occurred later in that period and took place most frequently among firms involved in oil exploration and production. Industry officials suggest that mergers increase efficiency, reduce costs, and enhance a company's ability to control prices.

Another reason for the drop in the number of U.S. refineries is that some OPEC countries have begun to refine their own oil products to maximize their profits. This strategy, employed particularly by Saudi Arabia, led to a drop in demand for U.S. refineries. As of November 2002 no major refinery manufacturer had plans to begin construction, a process that takes three to five years to complete. The last large refinery built in the United States was completed in 1976, and the last completed refinery of any size began operation in Valdez, Alaska, in 1993. The *Annual Energy Outlook 2004*, produced by the Energy Information Administration (EIA), stated that financial and legal considerations make it unlikely that new refineries will be built in the United States.

LOSS AND VOLATILITY OF OIL INDUSTRY JOBS

As mergers in the oil industry restructure companies and make combined operations more efficient, not only are there fewer refineries, there are fewer jobs. From 1982 to 1992, for example, there was a loss of 28% of petroleum refining jobs.

In 1996, after a decade of low oil prices, drilling slowed and the demand for rigs collapsed. New rig construction stopped altogether. Thousands of rigs were left idle, sold for scrap metal, or shipped overseas, and their crews were put out of work. Idle rigs became a source of spare parts for those still operating. In 1997, following a rise in oil prices, the demand for rigs soared, but by 1998 the market for rigs had once again dwindled as oil prices sank. Rising oil prices resulting from OPEC restrictions in 1999 eventually boosted the demand for drilling equipment. Although crude oil prices declined in 2001, they rose again in 2003 and reached an all-time high in late 2004, again spurring demand for oil rigs. According to Baker Hughes Inc., a company that has tallied weekly U.S. drilling activity since 1940, domestic oil drilling has rebounded sharply since late April 1999, when a low point was reached following the oil price collapse of late 1997.

Job loss has also occurred in other facets of the oil industry. The number of seismic land crews and marine vessels searching for oil in the United States and its waters decreased sharply after 1981. From 1982 to 1992, oil-extraction companies lost 51% of their workforce. In Texas, once the center of the U.S. oil industry, jobs plummeted from 80,000 in 1981 to 25,000 in 1996. In 1998 the Texas Comptroller of Public Accounts estimated that for every $1 drop in the price of oil per barrel, 10,000 jobs are lost in the Texas economy. That translated into 100,000 jobs lost in the Texas oil industry from October 1997 to December 1998. According to a February 6, 1999, article in the *New York Times* ("Oil Industry Sees More Losses"), the oil industry lost 24,415 jobs between November 1997 and February 1999 because of the decline of oil prices during that period. But domestic oil drilling rebounded sharply between mid-1999 and late 2004, creating an upswing in numbers of oil industry jobs.

DOMESTIC PRODUCTION

U.S. production of petroleum reached its highest level in 1970 at 11.3 million barrels per day total. (See Table 2.2.) Of that amount, 9.6 million barrels per day were crude oil. (See Figure 2.4.) After 1970 domestic production of petroleum declined. By 2003 U.S. domestic production averaged about 7.5 million barrels per day. (See Table 2.2.) Of that amount, 5.7 million barrels per day were crude oil. (See Figure 2.4.) Figure 2.5 shows the overall flow of petroleum in the United States for 2003.

According to the *Annual Energy Review 2003*, published by the EIA, in 2003 the 520,000 producing wells in the United States produced an average of eleven barrels per day per well, significantly below peak levels of more than eighteen barrels per day per well in the early 1970s. Any new oil discoveries are unlikely to result in a significant increase in domestic production in the near future because of the long lead-time needed to prepare for production.

Most domestic oil production takes place in only a few U.S. states. Texas, Alaska, Louisiana, California, and the offshore areas around these states produce about 75% of the nation's oil. Most domestic oil (about 64%, or 3.7

TABLE 2.2

Petroleum production, selected years, 1949–2003

(Thousand barrels per day)

Year	Production — Crude oil: 48 states[1]	Production — Crude oil: Alaska	Production — Crude oil: Total	Production — Natural gas plant liquids	Production — Total	Other domestic supply[2]	Trade — Imports	Trade — Exports	Trade — Net imports	Stock change[3]	Crude oil losses and unaccounted for[4]	Petroleum products supplied
1949	5,046	0	5,046	430	5,447	−2	645	327	318	−8	38	5,763
1950	5,407	0	5,407	499	5,906	2	850	305	545	−56	51	6,458
1955	6,807	0	6,807	771	7,578	34	1,248	368	880	(s)	37	8,455
1960	7,034	2	7,035	929	7,965	146	1,815	202	1,613	−83	8	9,797
1965	7,774	30	7,804	1,210	9,014	220	2,468	187	2,281	−8	10	11,512
1970	9,408	229	9,637	1,660	11,297	359	3,419	259	3,161	103	16	14,697
1971	9,245	218	9,463	1,693	11,155	382	3,926	224	3,701	71	−45	15,212
1972	9,242	199	9,441	1,744	11,185	388	4,741	222	4,519	−232	−43	16,367
1973	9,010	198	9,208	1,738	10,946	483	6,256	231	6,025	135	11	17,308
1974	8,581	193	8,774	1,688	10,462	516	6,112	211	5,862	179	38	16,653
1975	8,183	191	8,375	1,633	10,008	497	6,056	209	5,846	32	−3	16,322
1976	7,958	173	8,132	1,604	9,736	515	7,313	233	7,090	−58	−63	17,461
1977	7,781	464	8,245	1,618	9,862	575	8,807	243	8,565	548	22	18,431
1978	7,478	1,229	8,707	1,567	10,275	549	8,363	362	8,002	−94	73	18,847
1979	7,151	1,401	8,552	1,584	10,135	571	8,456	471	7,985	173	6	18,513
1980	6,980	1,617	8,597	1,573	10,170	641	6,909	544	6,365	140	−20	17,056
1981	6,962	1,609	8,572	1,609	10,180	558	5,996	595	5,401	160	−78	16,058
1982	6,953	1,696	8,649	1,550	10,199	583	5,113	815	4,298	−147	−68	15,296
1983	6,974	1,714	8,688	1,559	10,246	541	5,051	739	4,312	−20	−112	15,231
1984	7,157	1,722	8,879	1,630	10,509	599	5,437	722	4,715	280	−183	15,726
1985	7,146	1,825	8,971	1,609	10,581	612	5,067	781	4,286	−103	−145	15,726
1986	6,814	1,867	8,680	1,551	10,231	674	6,224	785	5,439	202	−139	16,281
1987	6,387	1,962	8,349	1,595	9,944	703	6,678	764	5,914	41	−145	16,665
1988	6,123	2,017	8,140	1,625	9,765	708	7,402	815	6,587	−28	−196	17,283
1989	5,739	1,874	7,613	1,546	9,159	722	8,061	859	7,202	−43	−200	17,325
1990	5,582	1,773	7,355	1,559	8,914	763	8,018	857	7,161	107	−257	16,988
1991	5,618	1,798	7,417	1,659	9,076	807	7,627	1,001	6,626	−10	−195	16,714
1992	5,457	1,714	7,171	1,697	8,868	900	7,888	950	6,938	−68	−258	17,033
1993	5,264	1,582	6,847	1,736	8,582	1,020	8,620	1,003	7,618	151	−168	17,237
1994	5,103	1,559	6,662	1,727	8,388	1,025	8,996	942	8,054	15	−266	17,718
1995	5,076	1,484	6,560	1,762	8,322	1,078	8,835	949	7,886	−246	−193	17,725
1996	5,071	1,393	6,465	1,830	8,295	1,150	9,478	981	8,498	−151	−215	18,309
1997	5,156	1,296	6,452	1,817	8,269	1,192	10,162	1,003	9,158	143	−145	18,620
1998	5,077	1,175	6,252	1,759	8,011	1,267	10,708	945	9,764	239	−115	18,917
1999	4,832	1,050	5,881	1,850	7,731	1,262	10,852	940	9,912	−422	−191	19,519
2000	4,851	970	5,822	1,911	7,733	1,325	11,459	1,040	10,419	−69	−155	19,701
2001	4,839	963	5,801	1,868	7,670	1,287	11,871	971	10,900	325	−117	19,649
2002	R4,761	984	R5,746	R1,880	R7,626	R1,374	R11,530	R984	R10,546	R−105	R−110	R19,761
2003P	P4,763	974	5,737	1,717	7,454	1,384	12,254	1,017	11,237	45	−14	20,044

TABLE 2.2

Petroleum production, selected years, 1949–2003 [CONTINUED]

(Thousand barrels per day)

[1]United States excluding Alaska and Hawaii.
[2]Refinery processing gains (refinery production minus refinery inputs), and field production of finished motor gasoline, motor gasoline blending components, and other hydrocarbons and oxygenates.
[3]A negative number indicates a decrease in stocks and a positive number indicates an increase. Distillate stocks in the "Northeast Heating Oil Reserve" are not included.
[4]"Unaccounted for" represents the difference between crude oil supply and disposition.
R = Revised.
P = Preliminary.
(s) = Less than 500 barrels per day.
Web Pages: For data not shown for 1951–1969, see http://www.eia.doe.gov/emeu/aer/petro.html. For related information, see http://www.eia.doe.gov/oil_gas/petroleum/info_glance/petroleum.html.

SOURCE: "Table 5.1. Petroleum Overview, Selected Years, 1949–2003 (Thousand Barrels per Day)," in *Annual Energy Review 2003*, U.S. Department of Energy, Energy Information Administration, Office of Energy Markets and End Use, September 7, 2004, http://www.eia.doe.gov/emeu/aer/pdf/aer.pdf (accessed September 28, 2004)

FIGURE 2.4

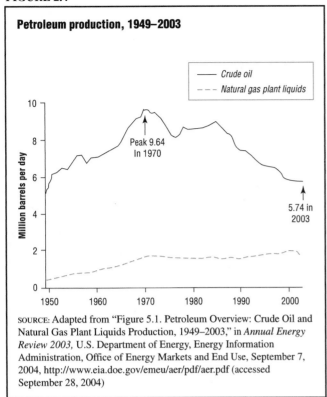

Petroleum production, 1949–2003

SOURCE: Adapted from "Figure 5.1. Petroleum Overview: Crude Oil and Natural Gas Plant Liquids Production, 1949–2003," in *Annual Energy Review 2003,* U.S. Department of Energy, Energy Information Administration, Office of Energy Markets and End Use, September 7, 2004, http://www.eia.doe.gov/emeu/aer/pdf/aer.pdf (accessed September 28, 2004)

million barrels per day) comes from onshore drilling, while the remaining 2.1 million barrels come from offshore sources. (See Figure 2.6.) Supplies from Alaska, which increased with the construction of a direct pipeline in the late 1970s, have begun to decline. Notice how the gap between "Total" and "48 States" is narrowing in Figure 2.7; this gap is Alaska's share of U.S. oil production.

Unless protected wildlife refuges in Alaska are opened for drilling, U.S. oil production will likely continue its decline. The Alaskan government and the administration of President George W. Bush strongly support drilling for oil in Alaska's Arctic National Wildlife Refuge. But there is great controversy over this proposal because of environmental concerns. The U.S. House of Representatives has approved drilling in the Arctic refuge in the past, only to have the proposal repeatedly fail in the Senate. As of November 2004 no agreement had been reached.

The United States is considered to be in a "mature" oil development phase, meaning that much of its oil has already been found. The amount of oil discovered per foot of exploratory well in the United States has fallen to less than half the rate of the early 1970s. Of the country's thirteen largest oil fields, seven are at least 80% depleted. Geological studies have estimated that 34% of the undiscovered recoverable resources are in Alaska, but it is uncertain whether the oil will ever be recovered.

Domestic production is also hampered by the expense of drilling and recovering oil in the United States compared to the expense incurred in Middle Eastern countries. Middle

Eastern producers can drill and bring out crude oil from enormous, easily accessible reservoirs for around $2 a barrel. In contrast, the U.S. Department of Energy (DOE) estimates it costs an American oil producer about $14 to produce a barrel of oil, not counting royalty payments and taxes, which add to the cost. Of all the successful domestic oil wells drilled, only about 1% have been "wildcat" wells that have led to the discovery of new large fields, and these discoveries have provided only minor additions to the total proved reserves.

DOMESTIC CONSUMPTION

In 2003 most petroleum was used for transportation (66%), followed by industrial use (25%), residential use (4%), electric utilities (3%), and commercial use (2%). (See Figure 2.8.)

Most petroleum used in the transportation sector is for motor gasoline. In the residential and commercial sectors, distillate fuel oil (refined fuels used for space heaters, diesel engines, and electric power generation) accounts for most petroleum use. Liquid petroleum gas (LPG) is the primary oil used in the industrial sector. In electric utilities residual fuel oils are used the most.

A modest decline in residual fuel oil consumption has been caused by the conversion of electric utilities and plants from heavy oil to coal or natural gas energy. An initial decline in the amount of motor gasoline used, beginning in 1978, was attributed to the federal Corporate Average Fuel Economy (CAFE) regulations, which required increased miles-per-gallon efficiency in new automobiles. However, motor gasoline use has increased steadily since then, partly from an increase in users and partly from a leveling off in vehicle efficiency as consumers once again prefer less efficient vehicles, such as larger automobiles and sport utility vehicles (SUVs).

WORLD OIL PRODUCTION AND CONSUMPTION

Total world petroleum production has increased somewhat steadily, reaching 69.5 million barrels per day in 2003, after a downturn in the early 1980s. (See Table 2.3.) The major producers in 2003 were, from most to least, Saudi Arabia, Russia, the United States, Iran, China, Mexico, Norway, United Arab Emirates, Venezuela and Canada, and Nigeria and Kuwait. (See Figure 2.9.) Together, Saudi Arabia, Russia, and the United States accounted for 33% of the world's crude oil production.

Like total world petroleum production, total world petroleum consumption has increased somewhat steadily, reaching 78.2 million barrels per day in 2002. (See Table 2.4.) In 2002 the leading petroleum consumers were, from most to least, the United States, Japan, China, and Russia. Figure 2.10 shows the major petroleum consumers of selected Organization for Economic Cooperation and Development (OECD) countries—a group of thirty nations

FIGURE 2.5

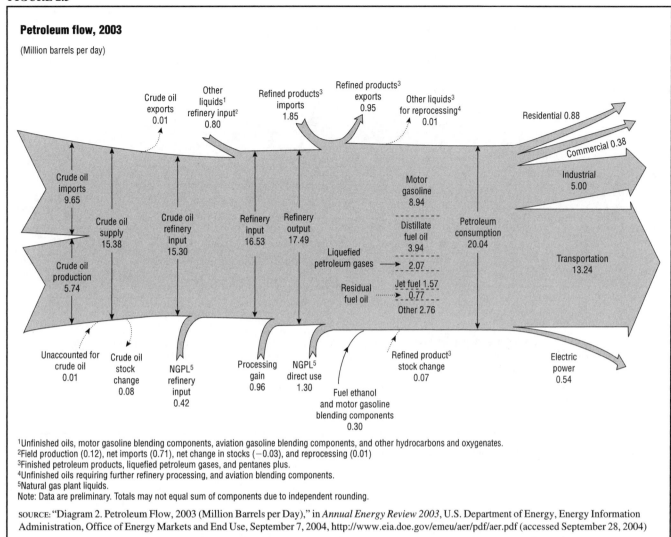

Petroleum flow, 2003

(Million barrels per day)

[1]Unfinished oils, motor gasoline blending components, aviation gasoline blending components, and other hydrocarbons and oxygenates.
[2]Field production (0.12), net imports (0.71), net change in stocks (−0.03), and reprocessing (0.01)
[3]Finished petroleum products, liquefied petroleum gases, and pentanes plus.
[4]Unfinished oils requiring further refinery processing, and aviation blending components.
[5]Natural gas plant liquids.
Note: Data are preliminary. Totals may not equal sum of components due to independent rounding.

SOURCE: "Diagram 2. Petroleum Flow, 2003 (Million Barrels per Day)," in *Annual Energy Review 2003*, U.S. Department of Energy, Energy Information Administration, Office of Energy Markets and End Use, September 7, 2004, http://www.eia.doe.gov/emeu/aer/pdf/aer.pdf (accessed September 28, 2004)

committed to democratic government and a market economy. They develop and refine economic and social policies. The major OECD consumers in 2002 were, in order, the United States, Japan, Germany, South Korea, Canada, France, Mexico, Italy, United Kingdom, and Spain. The United States was by far the leading consumer in either category, using 19.8 million barrels per day, followed by Japan (5.3 million barrels a day), China (5.2 million barrels a day), Germany (2.7 million barrels a day), and Russia (2.6 million barrels a day). (See Table 2.4.)

OIL IMPORTS AND EXPORTS

Around the world there are inequities between those countries that use petroleum and those that possess it. Countries with surpluses (Saudi Arabia, for example) sell their excess to others that need more than they can produce (the United States, China, Japan, and western European countries). Petroleum is sold as crude oil or as refined products. World trade has been moving toward refined products, as the petroleum-exporting countries

have realized that they can make a greater profit from refined oil products than from crude oil.

Though the United States produces a significant amount of petroleum, it has been importing oil since World War II (1939–45). This reflects the gradual exhaustion of reserves in the United States and the growing energy demand caused by population growth and economic expansion. Initially, the relatively low price of foreign oil encouraged U.S. dependence on it. American industry and economic life have been built on oil's cheap availability. But the price of foreign oil has gone up, and OPEC controls its availability.

Relatively low crude oil prices and the resulting reduced domestic oil production are the major causes of an increase in imports since 1985. From a low total net import (imports minus exports) of 4.3 million barrels per day in 1985, oil net imports increased to 11.2 million barrels per day by 2003. (See Table 2.2.) In 1985 imported oil supplied only 27.3% of American oil consumption. Just five years later, in 1990, the proportion had risen to 42%, and by 2003 it was 56.1% as demand continued to grow. (See

FIGURE 2.6

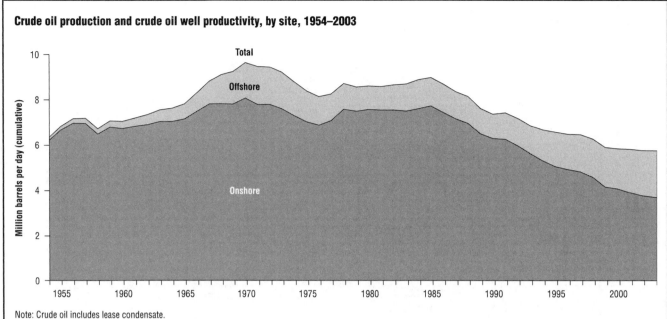

Crude oil production and crude oil well productivity, by site, 1954–2003

Note: Crude oil includes lease condensate.

SOURCE: Adapted from "Figure 5.2. Crude Oil Production and Crude Oil Well Productivity, 1954–2003: By Site," in *Annual Energy Review 2003*, U.S. Department of Energy, Energy Information Administration, Office of Energy Markets and End Use, September 7, 2004, http://www.eia.doe.gov/emeu/aer/pdf/aer.pdf (accessed September 28, 2004)

FIGURE 2.7

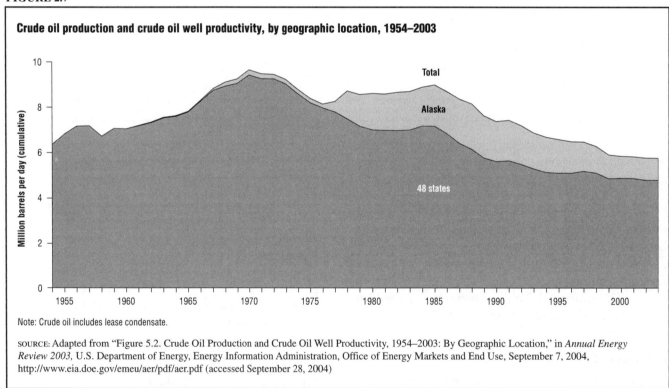

Crude oil production and crude oil well productivity, by geographic location, 1954–2003

Note: Crude oil includes lease condensate.

SOURCE: Adapted from "Figure 5.2. Crude Oil Production and Crude Oil Well Productivity, 1954–2003: By Geographic Location," in *Annual Energy Review 2003*, U.S. Department of Energy, Energy Information Administration, Office of Energy Markets and End Use, September 7, 2004, http://www.eia.doe.gov/emeu/aer/pdf/aer.pdf (accessed September 28, 2004)

Figure 1.6 in Chapter 1.) According to the Energy Information Administration's *Annual Energy Review 2003*, the leading suppliers of petroleum to the United States in 2003 were, from most to least, Canada, Saudi Arabia, Mexico, Venezuela, Nigeria, Iraq, United Kingdom, and Norway.

CONCERN ABOUT OIL DEPENDENCY

In the 1970s the United States and its leaders were very concerned that so much of the U.S. economic structure, based heavily on oil, was dependent upon the decisions of OPEC countries. Oil resources became an issue of national

FIGURE 2.8

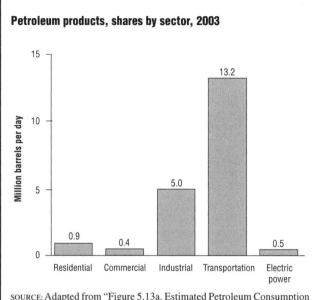

Petroleum products, shares by sector, 2003

SOURCE: Adapted from "Figure 5.13a. Estimated Petroleum Consumption by Sector: End Use and Electric Power Shares, 1949 and 2003," in *Annual Energy Review 2003*, U.S. Department of Energy, Energy Information Administration, Office of Energy Markets and End Use, September 7, 2004, http://www.eia.doe.gov/emeu/aer/pdf/aer.pdf (accessed September 28, 2004)

security, and OPEC countries, especially the Arab members, were often portrayed as potentially strangling the U.S. economy. The decisions of the Ronald Reagan and George H. W. Bush administrations in the 1980s and early 1990s to permit the energy issue to be handled by the marketplace was consistent with their economic philosophy but indicated that they saw oil supply as an economic, not a political, issue. This downplayed the political side of the energy problem in the international arena. The Clinton administration was unable to do much about America's dependence on foreign oil, as low prices throughout most of the 1990s set back energy conservation measures and public concern. In the United States efficiency gains in automobiles have been offset by the preference for large vehicles, such as SUVs. Such preferences imply lack of public interest in reducing the consumption of foreign oil. By 2003 about 63% of the nation's crude oil supply came from outside the country, as shown in Figure 2.5, and 42.2% of that came from OPEC nations, as reported in the *Annual Energy Review 2003*.

The decline in public concern about America's dependence on foreign oil is the result of factors other than low prices and the desire for large vehicles. One factor is a "comfort level" with oil that has been achieved through oil reserves, non-OPEC suppliers, and decreased demand. The United States and European nations have developed substantial reserves to withstand oil supply fluctuations. These reserves include the Strategic Petroleum Reserve (SPR) in the United States and government-required reserves in Europe. Furthermore, in emergencies, non-OPEC oil producers such as the United Kingdom and Norway can

increase their output. Although demand for oil had stabilized in the 1990s because of the conservation efforts of many industrialized nations, particularly European countries, the demand increased in the early 2000s as China, Japan, and the United States consumed more energy.

The increased use of pipelines across Saudi Arabia and Turkey has made the job of picking up oil from these countries safer. A growing number of tankers pick up their oil in either the Red Sea or the Mediterranean Sea before delivering it to Europe or the United States. These ships do not have to go through the potentially dangerous Persian Gulf. In addition, since many American strategists are wary of navigation in the Straits of Hormuz, where a future enemy might be able to stop the flow of oil to the West, shipment through pipelines lessens the importance of the waterway. On the other hand, such pipelines could be destroyed relatively easily in a war.

Based on these factors, if the oil shortages that developed in 1973 and 1979 occurred again, the result would likely not be the same. Despite the U.S. role as protector of Kuwait's oil in the Persian Gulf, there seemed to be little serious concern that America's dependence on foreign oil, especially OPEC oil, represented a threat to national security or national stability. This position changed somewhat after the terrorist attacks of September 11, 2001, and their aftermath—the "War on Terrorism," which included the war with Iraq and the consequent unrest in Middle Eastern nations. These events resulted in a sense that the United States should be less dependent on foreign oil, yet demand for the product persisted. In 2002 experts suggested that if the United States attacked Iraq, as it ultimately did in March 2003, the result might be a stabilization of the supply and price of oil (Michael E. Kanell, "War in Iraq: Is It about Oil?" *The [Montreal] Gazette,* October 26, 2002). Nevertheless, in late 2004 oil prices had not stabilized. In fact, oil prices increased dramatically at that time as concerns about terrorism in Iraq and elsewhere rose, the value of the dollar declined, demand for oil remained high, and oil supplies and reserves were tight.

PROJECTED OIL SUPPLY AND CONSUMPTION

The EIA, in its *Annual Energy Outlook 2004,* projected that domestic crude oil production in the lower forty-eight states would increase from 4.6 million barrels per day in 2002 to 5.2 million barrels per day in 2008, then decline to 4.1 million barrels per day in 2025. The projected peak in production in 2008 is attributed to offshore production. The EIA projects that total offshore oil production will rise to 2.5 million barrels per day in 2008, then decline to 2.1 million barrels per day in 2025. Onshore lower forty-eight oil production is projected to decline, with 2025 values ranging from 1.9 million barrels per day to 2.1 million barrels per day. Crude oil production in Alaska is expected to continue at about 900,000

TABLE 2.3

World crude oil production, 1960–2003

(Million barrels per day)

Year	Persian Gulf nations[2]	Selected OPEC[1] producers Iran	Iraq	Kuwait[3]	Nigeria	Saudi Arabia[3]	United Arab Emirates	Venezuela	Total OPEC	Selected non-OPEC producers Canada	China	Mexico	Norway	Former U.S.S.R.	Russia	United Kingdom	United States	Total non-OPEC[4]	World
1960	5.27	1.07	0.97	1.69	0.02	1.31	0.00	2.85	8.70	0.52	0.10	0.27	0.00	2.91	—	(s)	7.04	12.29	20.99
1961	5.65	1.20	1.01	1.74	0.05	1.48	0.00	2.92	9.36	0.61	0.11	0.29	0.00	3.28	—	(s)	7.18	13.09	22.45
1962	6.19	1.33	1.01	1.96	0.07	1.64	0.01	3.20	10.51	0.67	0.12	0.31	0.00	3.67	—	(s)	7.33	13.84	24.35
1963	6.82	1.49	1.16	2.10	0.08	1.79	0.05	3.25	11.51	0.71	0.13	0.31	0.00	4.07	—	(s)	7.54	14.62	26.13
1964	7.61	1.71	1.26	2.30	0.12	1.90	0.19	3.39	12.98	0.75	0.18	0.32	0.00	4.60	—	(s)	7.61	15.20	28.18
1965	8.37	1.91	1.32	2.36	0.27	2.21	0.28	3.47	14.35	0.81	0.23	0.32	0.00	4.79	—	(s)	7.80	15.98	30.33
1966	9.32	2.13	1.39	2.48	0.42	2.60	0.36	3.37	15.77	0.88	0.29	0.33	0.00	5.23	—	(s)	8.30	17.19	32.96
1967	9.91	2.60	1.23	2.50	0.32	2.81	0.38	3.54	16.85	0.96	0.28	0.36	0.00	5.68	—	(s)	8.81	18.54	35.39
1968	10.91	2.84	1.50	2.61	0.14	3.04	0.50	3.60	18.79	1.19	0.30	0.39	0.00	6.08	—	(s)	9.10	19.84	38.63
1969	11.95	3.38	1.52	2.77	0.54	3.22	0.63	3.59	20.91	1.13	0.48	0.46	0.00	6.48	—	(s)	9.24	20.79	41.70
1970	13.39	3.83	1.55	2.99	1.08	3.80	0.78	3.71	23.30	1.26	0.60	0.49	0.00	6.99	—	(s)	9.64	22.59	45.89
1971	15.77	4.54	1.69	3.20	1.53	4.77	1.06	3.55	25.21	1.35	0.78	0.49	0.01	7.48	—	(s)	9.46	23.31	48.52
1972	17.54	5.02	1.47	3.28	1.82	6.02	1.20	3.22	26.89	1.53	0.90	0.51	0.03	7.89	—	(s)	9.44	24.25	51.14
1973	20.67	5.86	2.02	3.02	2.05	7.60	1.53	3.37	30.63	1.80	1.09	0.47	0.03	8.32	—	(s)	9.21	25.05	55.68
1974	21.28	6.02	1.97	2.55	2.26	8.48	1.68	2.98	30.35	1.55	1.32	0.57	0.04	8.91	—	(s)	8.77	25.37	55.72
1975	18.93	5.35	2.26	2.08	1.78	7.08	1.66	2.35	26.77	1.43	1.49	0.71	0.19	9.52	—	0.01	8.37	26.06	52.83
1976	21.51	5.88	2.42	2.15	2.07	8.58	1.94	2.29	30.33	1.31	1.67	0.83	0.28	10.06	—	0.25	8.13	27.01	57.34
1977	21.73	5.66	2.35	1.97	2.09	9.25	2.00	2.24	30.89	1.32	1.87	0.98	0.28	10.60	—	0.77	8.24	28.82	59.71
1978	20.61	5.24	2.56	2.13	1.90	8.30	1.83	2.17	29.46	1.32	2.08	1.21	0.36	11.11	—	1.08	8.71	30.70	60.16
1979	21.07	3.17	3.48	2.50	2.30	9.53	1.83	2.36	30.58	1.50	2.12	1.46	0.40	11.38	—	1.57	8.55	32.09	62.67
1980	17.96	1.66	2.51	1.66	2.06	9.90	1.71	2.17	26.61	1.44	2.11	1.94	0.53	11.71	—	1.62	8.60	32.99	59.60
1981	15.25	1.38	1.00	1.13	1.43	9.82	1.47	2.10	22.48	1.29	2.01	2.31	0.50	11.85	—	1.81	8.57	33.60	56.08
1982	12.16	2.21	1.01	0.82	1.30	6.48	1.25	1.90	18.78	1.27	2.05	2.75	0.52	11.91	—	2.07	8.65	34.70	53.48
1983	11.08	2.44	1.01	1.06	1.24	5.09	1.15	1.80	17.50	1.36	2.12	2.69	0.61	11.97	—	2.29	8.69	35.76	53.26
1984	10.78	2.17	1.21	1.16	1.39	4.66	1.15	1.80	17.44	1.44	2.30	2.78	0.70	11.86	—	2.48	8.88	37.05	54.49
1985	9.63	2.25	1.43	1.02	1.50	3.39	1.19	1.68	16.18	1.47	2.51	2.75	0.79	11.59	—	2.53	8.97	37.80	53.98
1986	11.70	2.04	1.69	1.42	1.47	4.87	1.33	1.79	18.28	1.47	2.62	2.44	0.87	11.90	—	2.54	8.68	37.95	56.23
1987	12.10	2.30	2.08	1.59	1.34	4.27	1.54	1.75	18.52	1.54	2.69	2.55	1.02	12.05	—	2.41	8.35	38.15	56.67
1988	13.46	2.24	2.69	1.49	1.45	5.09	1.57	1.90	20.32	1.62	2.73	2.51	1.16	12.05	—	2.23	8.14	38.42	58.74
1989	14.84	2.81	2.90	1.78	1.72	5.06	1.86	1.91	22.07	1.56	2.76	2.52	1.55	11.72	—	1.80	7.61	37.79	59.86
1990	15.28	3.09	2.04	1.18	1.81	6.41	2.12	2.14	23.20	1.55	2.77	2.55	1.70	10.98	—	1.82	7.36	37.37	60.57
1991	14.74	3.31	0.31	0.19	1.89	8.12	2.39	2.38	23.27	1.55	2.84	2.68	1.89	9.99	—	1.80	7.42	36.94	60.21
1992	15.97	3.43	0.43	1.06	1.94	8.33	2.27	2.37	24.40	1.61	2.85	2.67	2.23	—	7.63	1.83	7.17	35.81	60.21
1993	16.71	3.54	0.51	1.85	1.96	8.20	2.16	2.45	25.12	1.68	2.89	2.67	2.35	—	6.73	1.92	6.85	35.12	60.24
1994	16.96	3.62	0.55	2.03	1.93	8.12	2.19	2.59	25.51	1.75	2.94	2.69	2.52	—	6.14	2.37	6.66	35.48	60.99
1995	17.21	3.64	0.56	2.06	1.99	8.23	2.23	2.75	26.00	1.81	2.99	2.62	2.77	—	6.00	2.49	6.56	36.33	62.33
1996	17.37	3.69	0.58	2.06	2.00	8.22	2.28	2.94	26.46	1.84	3.13	2.86	3.10	—	5.85	2.57	6.46	37.25	63.71
1997	18.10	3.66	1.16	2.01	2.13	8.36	2.32	3.28	27.71	1.92	3.20	3.02	3.14	—	5.92	2.52	6.45	37.98	65.69
1998	19.34	3.63	2.15	2.09	2.15	8.39	2.35	3.17	28.77	1.98	3.20	3.07	3.02	—	5.85	2.62	6.25	38.15	66.92
1999	18.67	3.56	2.51	1.09	2.13	7.83	2.17	2.83	27.58	1.91	3.20	2.91	3.02	—	6.08	2.68	5.88	38.27	65.85
2000	19.89	3.70	2.57	2.08	2.17	8.40	2.37	3.16	29.26	1.98	3.25	3.01	3.20	—	6.48	2.28	5.80	39.08	68.34
2001	[R]19.10	3.72	[R]2.39	2.00	2.26	8.03	[R]2.21	[R]3.01	[R]28.34	2.03	3.30	3.16	3.12	—	[R]6.92	2.28	5.80	[R]39.60	[R]67.94
2002	17.79	3.44	2.02	1.89	2.12	7.63	2.08	[R]2.60	26.37	2.17	3.39	3.18	2.99	—	7.41	2.29	[R]5.75	[R]40.47	[R]66.84
2003[P]	19.26	3.78	1.31	2.18	2.24	8.85	2.35	2.34	28.01	2.31	3.41	3.37	2.85	—	8.18	2.09	5.74	41.49	69.50

TABLE 2.3

World crude oil production, 1960–2003 [CONTINUED]

(Million barrels per day)

[1]Organization of Petroleum Exporting Countries.
[2]Persian Gulf Nations are Bahrain, Iran, Iraq, Kuwait, Qatar, Saudi Arabia, and United Arab Emirates.
[3]Includes about one-half of the production in the Neutral Zone between Kuwait and Saudi Arabia.
[4]Ecuador, which withdrew from OPEC on December 31, 1992, and Gabon, which withdrew on December 31, 1994, are included in "Non-OPEC" for all years.
R = Revised.
P = Preliminary.
— = Not applicable.
(s) = Less than 0.005 million barrels per day.
Notes: Includes lease condensate, excludes natural gas plant liquids. Totals may not equal sum of components due to independent rounding.
Web Page: For related information, see http://www.eia.doe.gov/international.

SOURCE: "Table 11.5. World Crude Oil Production, 1960–2003 (Million Barrels per Day)," in *Annual Energy Review 2003*, U.S. Department of Energy, Energy Information Administration, Office of Energy Markets and End Use, September 7, 2004, http://www.eia.doe.gov/emeu/aer/pdf/aer.pdf (accessed September 28, 2004)

FIGURE 2.9

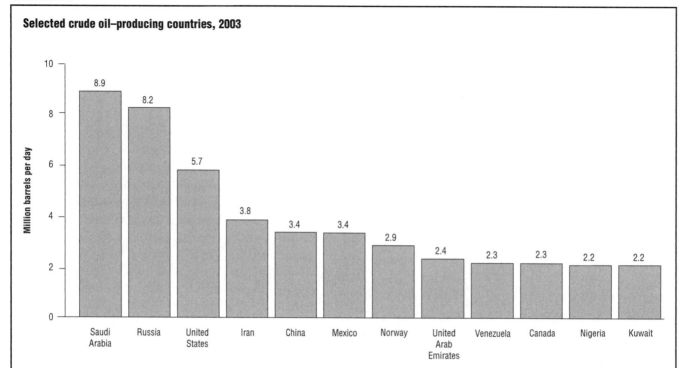

Selected crude oil–producing countries, 2003

SOURCE: Adapted from "Figure 11.5. World Crude Oil Production: Selected Producing Countries, 2003," in *Annual Energy Review 2003*, U.S. Department of Energy, Energy Information Administration, Office of Energy Markets and End Use, September 7, 2004, http://www.eia.doe.gov/emeu/aer/pdf/aer.pdf (accessed September 28, 2004)

barrels per day through 2016; it is then projected to decline to 510,000 barrels per day in 2025.

Due to declines in crude oil production after 2008 and in Alaska in 2016, the domestic petroleum supply is expected to decline after 2008 as well. However, consumption is projected to increase significantly from 2002 to 2025 (see Figure 2.11.) Thus, the EIA projects an increasing dependence on petroleum imports, with imports filling 70% of American petroleum needs in 2025, up from 53% in 2002. (See Figure 2.11.)

STRATEGIC PETROLEUM RESERVES

In 1923 the Warren G. Harding administration set up the Petroleum Reserve to ensure that the U.S. Navy would have adequate fuel in the event of war. In 1975, in response to the growing concern over America's energy dependence, Congress turned the Strategic Petroleum Reserve (SPR) over to the Department of the Interior under the Energy Policy and Conservation Act (PL 94–163). An additional law, PL 101–383, expanded the SPR and created a second reserve for refined products.

SPR oil is stored in deep salt caverns located at four storage sites in Louisiana and two in Texas. (Oil does not dissolve salt the way water does.) Each site's caverns vary in capacity, but most can hold approximately 10 million barrels of oil. The SPR system currently has forty-one such caverns. If the United States suddenly found its sup-

plies cut off, the reserve system would be connected to existing commercial lines that would start pumping the oil.

At the end of 2003 the SPR contained 638 million barrels (see Figure 2.12), enough to equal fifty-seven days of imported oil should the supply be cut off (see Figure 2.13.) The SPR increased in 2002 and 2003 due to the importation of sixty-four million barrels of crude oil for this purpose. Figure 2.13 shows a decline in the reserves, from a high of 115 days in 1985 to the current fifty-seven days. This decline is the result of two major factors:

1) It reflects the increase in imports since 1985. As the nation has imported a greater amount of oil, the days of import replacement represented by the amount of oil in the SPR have dropped.

2) It reflects a real decline in quantity. In 1996–97, millions of barrels of oil were sold (at a loss of $10 a barrel or so) to help balance the federal budget. Also, oil was withdrawn from the reserves at the start of the Persian Gulf War in 1990–91. In September 2000 President Clinton authorized the release of 30 million barrels of oil to bolster oil supplies, particularly heating oil in the Northeast.

OIL PRICES

The law of supply and demand usually explains oil price changes; the price of goods reflects a relationship

TABLE 2.4

World petroleum consumption, 1960–2002

(Million barrels per day)

Year	Canada	France	Germany[2]	Italy	Japan	Mexico[3]	South Korea[3]	Spain	United Kingdom	United States	Total OECD[4]	Brazil	China	India	Former U.S.S.R.	Russia	Total non-OECD	World
1960	0.84	0.56	0.63	0.44	0.66	0.30	0.01	0.10	0.94	9.80	15.78	0.27	0.17	0.16	2.38	—	5.56	21.34
1961	0.87	0.63	0.79	0.54	0.82	0.29	0.02	0.12	1.04	9.98	16.77	0.28	0.17	0.17	2.57	—	6.23	23.00
1962	0.92	0.73	1.00	0.67	0.93	0.30	0.02	0.12	1.12	10.40	18.06	0.31	0.14	0.18	2.87	—	6.83	24.89
1963	0.99	0.86	1.17	0.77	1.21	0.31	0.03	0.12	1.27	10.74	19.60	0.34	0.17	0.21	3.15	—	7.32	26.92
1964	1.05	0.98	1.36	0.90	1.48	0.33	0.02	0.20	1.36	11.02	21.05	0.35	0.20	0.22	3.58	—	8.03	29.08
1965	1.14	1.09	1.61	0.98	1.74	0.34	0.03	0.23	1.49	11.51	22.81	0.33	0.23	0.25	3.61	—	8.33	31.14
1966	1.21	1.19	1.80	1.08	1.98	0.36	0.04	0.31	1.58	12.08	24.60	0.38	0.30	0.28	3.87	—	8.96	33.56
1967	1.25	1.34	1.86	1.19	2.14	0.39	0.07	0.36	1.64	12.56	25.94	0.38	0.28	0.26	4.22	—	9.65	35.59
1968	1.34	1.46	1.99	1.40	2.66	0.41	0.10	0.46	1.82	13.39	28.56	0.46	0.31	0.31	4.48	—	10.40	38.96
1969	1.42	1.66	2.33	1.69	3.25	0.45	0.15	0.49	1.98	14.14	31.54	0.48	0.44	0.34	4.87	—	11.35	42.89
1970	1.52	1.94	2.83	1.71	3.82	0.50	0.20	0.58	2.10	14.70	34.49	0.53	0.62	0.40	5.31	—	12.32	46.81
1971	1.56	2.12	2.94	1.84	4.14	0.52	0.23	0.64	2.14	15.21	36.07	0.58	0.79	0.42	5.66	—	13.35	49.42
1972	1.66	2.32	3.13	1.95	4.36	0.59	0.23	0.68	2.28	16.37	38.74	0.66	0.91	0.46	6.12	—	14.35	53.09
1973	1.73	2.60	3.34	2.07	4.95	0.67	0.28	0.78	2.34	17.31	41.53	0.78	1.12	0.49	6.60	—	15.71	57.24
1974	1.78	2.45	3.06	2.00	4.86	0.71	0.29	0.86	2.21	16.65	40.12	0.86	1.19	0.47	7.28	—	16.56	56.68
1975	1.78	2.25	2.96	1.86	4.62	0.75	0.31	0.87	1.91	16.32	38.82	0.92	1.36	0.50	7.52	—	17.38	56.20
1976	1.82	2.42	3.21	1.97	4.84	0.83	0.36	0.97	1.89	17.46	41.39	1.00	1.53	0.51	7.78	—	18.28	59.67
1977	1.85	2.29	3.21	1.90	4.88	0.88	0.42	0.94	1.91	18.43	42.43	1.02	1.64	0.55	8.18	—	19.40	61.83
1978	1.90	2.41	3.29	1.95	4.95	0.99	0.48	0.98	1.94	18.85	43.62	1.11	1.79	0.62	8.48	—	20.54	64.16
1979	1.97	2.46	3.37	2.04	5.05	1.10	0.53	1.02	1.97	18.51	44.01	1.18	1.84	0.66	8.64	—	21.21	65.22
1980	1.87	2.26	3.08	1.93	4.96	1.27	0.54	0.99	1.73	17.06	R41.76	1.15	1.77	0.64	9.00	—	R21.35	R63.11
1981	1.77	2.02	2.80	1.87	4.85	1.40	0.54	0.94	1.59	16.06	R39.49	1.09	1.71	0.73	8.94	—	R21.45	R60.94
1982	1.58	1.88	2.74	1.78	4.58	1.48	0.53	1.00	1.59	15.30	R37.77	1.06	1.66	0.74	9.08	—	R21.77	R59.54
1983	1.45	1.84	2.66	1.75	4.40	1.35	0.56	1.01	1.53	15.23	R36.91	0.98	1.73	0.77	8.95	—	R21.87	R58.78
1984	R1.52	R1.77	R2.56	R1.72	R4.67	R1.40	R0.55	R0.85	R1.83	15.73	R37.70	1.03	1.74	0.82	8.91	—	R22.13	59.83
1985	R1.53	R1.75	R2.65	R1.71	R4.44	R1.48	R0.55	R0.86	R1.62	15.73	R37.48	1.08	1.89	R0.89	8.95	—	R22.61	60.09
1986	R1.53	R1.76	R2.79	R1.73	R4.50	R1.52	R0.59	R0.87	R1.64	16.28	R38.61	1.24	2.00	0.95	8.98	—	R23.22	R61.83
1987	R1.61	R1.79	R2.72	R1.82	R4.57	R1.58	R0.63	0.90	R1.61	16.67	R39.37	1.26	2.12	0.99	9.00	—	R23.76	R63.13
1988	R1.68	R1.80	R2.72	R1.83	R4.85	R1.60	R0.75	0.98	R1.69	17.28	R40.68	1.30	2.28	1.08	8.89	—	R24.32	R65.00
1989	R1.75	R1.84	R2.58	R1.90	R5.06	R1.72	R0.86	0.98	R1.73	17.33	R41.34	1.32	2.38	1.15	8.74	—	R24.76	R66.10
1990	R1.75	R1.83	R2.68	1.87	R5.30	R1.75	R1.05	1.01	R1.78	16.99	R41.60	1.47	2.30	1.17	8.39	—	R24.93	R66.53
1991	R1.67	R1.94	2.83	1.86	R5.37	R1.83	R1.26	1.07	1.80	16.71	R41.97	1.48	2.50	1.19	8.35	—	R25.13	R67.10
1992	R1.73	R1.93	R2.84	R1.89	R5.49	R1.86	R1.53	R1.10	R1.82	17.03	R42.92	1.52	2.66	1.27	—	4.42	R24.32	R67.24
1993	R1.76	R1.88	R2.91	R1.89	R5.41	R1.84	R1.68	1.06	R1.83	17.24	R43.29	1.58	2.96	1.31	—	3.75	R24.11	R67.40
1994	R1.77	R1.87	2.88	R1.87	R5.70	R1.93	R1.84	R1.12	R1.83	17.72	R44.46	1.67	3.16	1.41	—	3.18	24.25	R68.71
1995	R1.81	R1.92	2.88	R1.94	R5.73	R1.82	2.01	R1.19	R1.81	17.72	R44.91	1.79	3.36	1.57	—	2.98	R25.09	R70.00
1996	R1.87	R1.95	R2.92	R1.92	R5.77	R1.79	R2.10	R1.20	R1.85	18.31	R45.98	1.90	3.61	1.68	—	2.62	R25.52	R71.50
1997	R1.95	R1.97	R2.92	R1.93	R5.72	R1.85	R2.25	R1.27	R1.80	18.62	R46.71	2.03	3.92	1.77	—	2.56	26.49	R73.20
1998	1.95	R2.04	R2.92	R1.94	R5.52	1.95	R1.92	R1.36	1.79	18.92	R46.87	2.10	4.11	1.84	—	2.49	R27.01	R73.88
1999	2.03	2.03	2.84	R1.89	R5.61	R1.96	R2.08	R1.40	R1.79	19.52	R47.76	2.13	4.36	2.03	—	2.54	R27.97	R75.73
2000	R2.02	R2.00	R2.77	R1.85	R5.48	R2.04	R2.14	R1.43	R1.76	19.70	R47.85	2.17	R4.80	2.13	—	2.58	R28.98	R76.83
2001	R2.04	R2.05	2.81	R1.84	R5.39	R1.99	R2.13	R1.49	R1.72	19.65	R47.90	R2.21	R4.92	R2.18	—	R2.74	R30.10	R78.00
2002[P]	2.09	1.98	2.72	1.85	5.30	1.98	2.18	1.51	1.70	19.76	47.82	2.16	5.16	2.19	—	2.58	30.39	78.21

TABLE 2.4

World petroleum consumption, 1960–2002 [CONTINUED]

(Million barrels per day)

[1]Organization for Economic Cooperation and Development.
[2]Through 1969, the data for Germany are for the former West Germany only. For 1970 through 1990, this is East and West Germany. Beginning in 1991, this is unified Germany.
[3]Mexico, which joined the OECD on May 18, 1994, and South Korea, which joined the OECD on December 12, 1996, are included in the OECD for all years shown in this table.
[4]Hungary and Poland, which joined the OECD on May 7, 1996, and November 22, 1996, respectively, are included in total OECD beginning in 1970, the first year that data for these countries were available. OECD totals include Czechoslovakia from 1980–1992, Czech Republic and Slovakia from 1992–2002.
R = Revised.
P = Preliminary.
— = Not applicable.
Note: Totals may not equal sum of components due to independent rounding.
Web Page: For related information, see http://www.eia.doe.gov/international.

SOURCE: "Figure 11.10. World Petroleum Consumption, 1960–2002 (Million Barrels per Day)," in *Annual Energy Review 2003*, U.S. Department of Energy, Energy Information Administration, Office of Energy Markets and End Use, September 7, 2004, http://www.eia.doe.gov/emeu/aer/pdf/aer.pdf (accessed September 28, 2004)

FIGURE 2.10

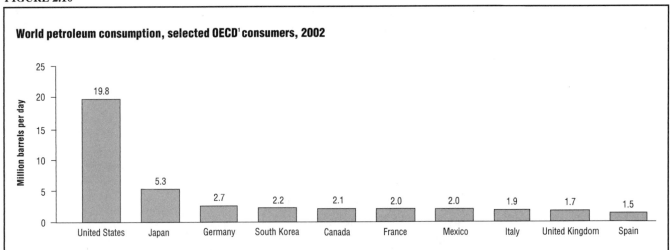

World petroleum consumption, selected OECD[1] consumers, 2002

[1]Organization for Economic Cooperation and Development.

SOURCE: Adapted from "Figure 11.10. World Petroleum Consumption: Selected OECD Consumers, 2002," in *Annual Energy Review 2003*, U.S. Department of Energy, Energy Information Administration, Office of Energy Markets and End Use, September 7, 2004, http://www.eia.doe.gov/emeu/aer/pdf/aer.pdf (accessed September 28, 2004)

FIGURE 2.11

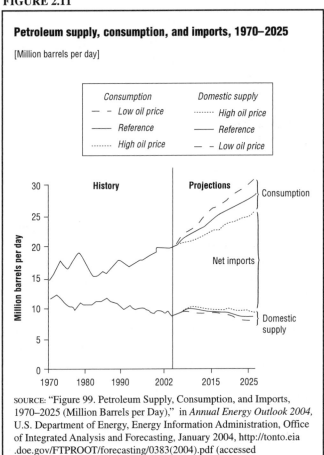

Petroleum supply, consumption, and imports, 1970–2025

[Million barrels per day]

SOURCE: "Figure 99. Petroleum Supply, Consumption, and Imports, 1970–2025 (Million Barrels per Day)," in *Annual Energy Outlook 2004*, U.S. Department of Energy, Energy Information Administration, Office of Integrated Analysis and Forecasting, January 2004, http://tonto.eia.doe.gov/FTPROOT/forecasting/0383(2004).pdf (accessed November 16, 2004)

FIGURE 2.12

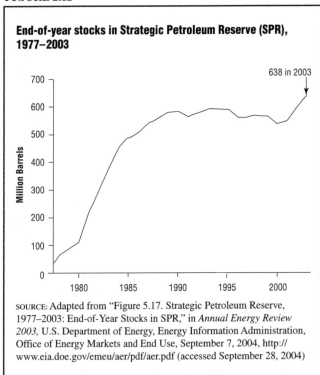

End-of-year stocks in Strategic Petroleum Reserve (SPR), 1977–2003

SOURCE: Adapted from "Figure 5.17. Strategic Petroleum Reserve, 1977–2003: End-of-Year Stocks in SPR," in *Annual Energy Review 2003*, U.S. Department of Energy, Energy Information Administration, Office of Energy Markets and End Use, September 7, 2004, http://www.eia.doe.gov/emeu/aer/pdf/aer.pdf (accessed September 28, 2004)

between the supply (availability) and the demand (need). Higher prices increase production, as it becomes profitable to operate more expensive wells, and reduce demand, as consumers lower usage and increase conser-vation efforts. The factors also work the other way: Reduced demand or increased supply generally cause the price of oil to drop.

The demand for petroleum products varies, and petro-leum prices usually fluctuate with demand. Heating oil demand rises during the winter. A cold spell, which leads to a sharp rise in demand, may result in a corresponding price increase. A warm winter may be reflected in lower prices as suppliers try to clear out their inventory. Gasoline

FIGURE 2.13

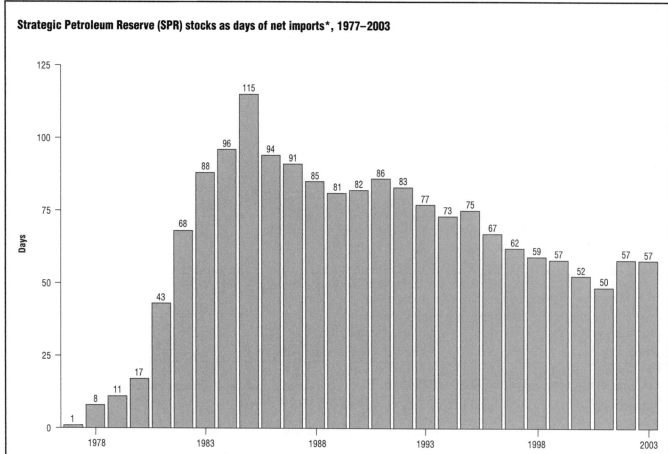

Strategic Petroleum Reserve (SPR) stocks as days of net imports*, 1977–2003

*Derived by dividing end-of-year SPR stocks by annual average daily net imports of all petroleum.

SOURCE: Adapted from "Figure 5.17. Strategic Petroleum Reserve, 1977–2003: SPR Stocks as Days' Worth of Net Imports," in *Annual Energy Review 2003,* U.S. Department of Energy, Energy Information Administration, Office of Energy Markets and End Use, September 7, 2004, http://www.eia.doe.gov/emeu/aer/pdf/aer.pdf (accessed September 28, 2004)

demand rises during the summer, when people drive more on vacations and for recreation, and gas prices usually rise as a consequence. Petroleum demand also reflects the general condition of the economy. During a recession, demand for and production of petroleum products drops. Wars and other types of political unrest in oil-producing nations add volatility to petroleum prices, which fluctuate—sometimes dramatically—depending on the situation at the time.

While consumers prefer low prices that allow them to save money or get more for the same price, producers naturally prefer to keep prices high. Oil producers formed the OPEC cartel in 1960. A cartel is a group of businesses that agree to control production and marketing to avoid competing with one another. Since 1973 OPEC has tried to control the oil supply in order to achieve higher prices.

OPEC has faced long-term problems, however, because high prices in the late 1970s to mid-1980s encouraged conservation, reducing demand for oil and leading to a sharp decline in oil prices. As a result of the decreased demand for oil and lower prices, OPEC lost some of its ability to control its members and, consequently, prices.

Nevertheless, OPEC actions can still effectively influence the petroleum market. For example, in an attempt to halt the downward slide of oil prices in 1999, Saudi Arabia, Mexico, and Venezuela agreed to cut production by 1.6% to 2 million barrels a day. Many other oil-producing nations also limited their production. The limitations worked: During the summer of 2000 oil prices climbed. According to a 2001 report by the Energy Information Administration (EIA) of the Department of Energy (*Annual Energy Review 2000*), the price in real dollars paid by refiners for crude oil in 1999 averaged $16.71 per barrel, and in 2000 averaged $26.40 per barrel.

Gasoline Prices

Many middle-aged and older Americans can remember when gas cost 30 cents per gallon in the early 1970s. From 1973 to 1981, the price of a gallon of leaded regular gasoline (in current dollars that do not consider inflation) more than tripled, while the price in real dollars (adjusted for inflation) rose 81%. (See Table 2.5.) In real dollars the price of a gallon of regular unleaded gasoline was $2.33 in 1981. However, after 1981, as a result of the international oil glut,

TABLE 2.5

Retail motor gasoline and on-highway diesel fuel prices, 1949–2003

(Dollars per gallon)

| | Motor gasoline by grade | | | | | | | | Regular motor gasoline by area type[1] | | | |
| | Leaded regular | | Unleaded regular | | Unleaded premium | | All grades | | Conventional gasoline areas[3,4] | Reformulated gasoline areas[5,6] | All areas | On-highway diesel fuel[1] |
Year	Nominal	Real[2]	Nominal	Real[2]	Nominal	Real[2]	Nominal	Real[2]				
1949	0.27	R1.64	NA	NA	NA	NA	NA	NA	NA	NA	NA	NA
1950	0.27	R1.62	NA	NA	NA	NA	NA	NA	NA	NA	NA	NA
1955	0.29	R1.55	NA	NA	NA	NA	NA	NA	NA	NA	NA	NA
1960	0.31	R1.48	NA	NA	NA	NA	NA	NA	NA	NA	NA	NA
1965	0.31	R1.39	NA	NA	NA	NA	NA	NA	NA	NA	NA	NA
1970	0.36	R1.30	NA	NA	NA	NA	NA	NA	NA	NA	NA	NA
1971	0.36	R1.26	NA	NA	NA	NA	NA	NA	NA	NA	NA	NA
1972	0.36	R1.20	NA	NA	NA	NA	NA	NA	NA	NA	NA	NA
1973	0.39	R1.22	NA	NA	NA	NA	NA	NA	NA	NA	NA	NA
1974	0.53	R1.53	NA	NA	NA	NA	NA	NA	NA	NA	NA	NA
1975	0.57	R1.49	NA	NA	NA	NA	NA	NA	NA	NA	NA	NA
1976	0.59	R1.47	0.61	R1.53	NA	NA	NA	NA	NA	NA	NA	NA
1977	0.62	R1.46	0.66	R1.53	NA	NA	NA	NA	NA	NA	NA	NA
1978	0.63	R1.37	0.67	R1.46	NA	NA	0.65	R1.43	NA	NA	NA	NA
1979	0.86	R1.73	0.90	R1.82	NA	NA	0.88	R1.78	NA	NA	NA	NA
1980	1.19	R2.20	1.25	R2.30	NA	NA	1.22	R2.26	NA	NA	NA	NA
1981	1.31	R2.22	1.38	R2.33	1.47	R2.49	1.35	R2.29	NA	NA	NA	NA
1982	1.22	R1.95	1.30	R2.07	1.42	R2.26	1.28	R2.04	NA	NA	NA	NA
1983	1.16	R1.77	1.24	R1.90	1.38	R2.12	1.23	R1.88	NA	NA	NA	NA
1984	1.13	R1.67	1.21	R1.79	1.37	R2.02	1.20	R1.77	NA	NA	NA	NA
1985	1.12	R1.60	1.20	R1.72	1.34	R1.92	1.20	R1.72	NA	NA	NA	NA
1986	0.86	R1.20	0.93	R1.30	1.09	R1.52	0.93	R1.31	NA	NA	NA	NA
1987	0.90	R1.23	0.95	R1.30	1.09	R1.49	0.96	R1.31	NA	NA	NA	NA
1988	0.90	R1.19	0.95	R1.25	1.11	R1.46	0.96	R1.27	NA	NA	NA	NA
1989	1.00	R1.27	1.02	R1.30	1.20	R1.52	1.06	R1.35	NA	NA	NA	NA
1990	1.15	R1.41	1.16	R1.43	1.35	R1.65	1.22	R1.49	NA	NA	NA	NA
1991	NA	NA	1.14	R1.35	1.32	R1.56	1.20	R1.42	1.10	NA	1.10	NA
1992	NA	NA	1.13	R1.31	1.32	R1.52	1.19	R1.38	1.09	NA	1.09	NA
1993	NA	NA	1.11	R1.25	1.30	R1.47	1.17	R1.33	41.07	NA	1.07	NA
1994	NA	NA	1.11	R1.23	1.31	R1.45	1.17	R1.30	1.07	NA	1.08	NA
1995	NA	NA	1.15	R1.25	1.34	R1.45	1.21	R131	1.10	61.16	1.11	1.11
1996	NA	NA	1.23	R1.31	1.41	R1.51	1.29	R1.37	1.19	1.28	1.22	1.24
1997	NA	NA	1.23	R1.29	1.42	R1.48	1.29	R1.35	1.19	1.25	1.20	1.20
1998	NA	NA	1.06	R1.10	1.25	R1.30	1.12	R1.16	1.02	1.08	1.03	1.04
1999	NA	NA	1.17	R1.19	1.36	R1.39	1.22	R1.25	1.12	1.20	1.14	1.12
2000	NA	NA	1.51	R1.51	1.69	R1.69	1.56	R1.56	1.46	1.54	1.48	1.49
2001	NA	NA	1.46	R1.43	1.66	R1.62	1.53	R1.50	1.38	1.50	1.42	1.40
2002	NA	NA	1.36	R1.31	R1.56	1.50	1.44	R1.39	1.31	1.41	1.35	1.32
2003	NA	NA	1.59	1.51	1.78	1.68	1.64	1.55	1.52	1.66	1.56	1.51

[1]Nominal dollars.

[2]In chained (2000) dollars, calculated by using gross domestic product implicit price deflators.

[3]Any area that does not require the sale of reformulated gasoline.

[4]For 1993–2000, data collected for oxygenated areas are included in "Conventional gasoline areas."

[5]"Reformulated Gasoline Areas" are ozone nonattainment areas designated by the Environmental Protection Agency that require the use of reformulated gasoline.

[6]For 1995–2000, data collected for combined oxygenated and reformulated areas are included in "Reformulated Gasoline Areas."

R = Revised.

NA = Not available.

Web Page: For data not shown for 1951–1969, see http://www.eia.doe.gov/emeu/aer/petro.html. For related information, see http://www.eia.doe.gov/oil_gas/petroleum/info_glance/petroleum.html.

SOURCE: "Table 5.24. Retail Motor Gasoline and On-Highway Diesel Fuel Prices, Selected Years, 1949–2003 (Dollars per Gallon)," in *Annual Energy Review 2003*, U.S. Department of Energy, Energy Information Administration, Office of Energy Markets and End Use, September 7, 2004, http://www.eia.doe.gov/emeu/aer/pdf/aer.pdf (accessed September 28, 2004)

---

real prices tumbled. In 1998 the price was only $1.10, and after price increases in 1999, the price of gas still only averaged $1.51 in 2000 and 2003. By the fall of 2004, however, gasoline prices hovered around $2.00 per gallon.

ENVIRONMENTAL CONCERNS ABOUT OIL TRANSPORTATION

Transporting oil carries significant environmental risks. According to the U.S. Department of the Interior, the cause of most transportation spills is oil tanker accidents, such as the grounding of the *Exxon Valdez* in 1989.

The *Exxon Valdez* Oil Spill

A number of events have influenced American attitudes toward oil production and use. One of the most notable occurred in March 1989, when the *Exxon Valdez* oil tanker hit a reef in Alaska, spilling 11 million gallons of crude oil into the waters of Prince William Sound. The cleanup cost Exxon $1.28 billion, which does not include legal costs or any valuation of the wildlife lost. The spill was an environmental disaster for a formerly pristine area. Even measures used to clean up the spill, such as washing the beaches with hot water, caused additional damage.

The *Exxon Valdez* spill also led to debate about added safety measures in the design of tankers. Tankers are bigger than ever before. In 1945 the largest tanker held 16,500 tons of oil; today, supertankers carry more than 550,000 tons. These supertankers are difficult to maneuver because of their size and are likely to spill more oil if damaged. Although there have been fewer spills since 1973, the amount of oil lost has been roughly the same.

The Oil Pollution Act of 1990

The *Exxon Valdez* oil spill led Congress to pass the Oil Pollution Act of 1990 (PL 101–380) after having debated the issue for sixteen years. The bill increased, but still limited, oil spillers' federal liability (financial responsibility) as long as a spill is not the result of "gross negligence." The bill also mandates compensation to those who are economically injured by oil-spill accidents. Damages that can be charged to oil companies are limited to $60 million for tanker accidents and $75 million for accidents at offshore facilities. The rest of the cleanup costs are paid from a $500 million oil-spill fund generated by a 1.3 cents-per-barrel tax on oil. Individual states still maintain the right to impose unlimited liability on spillers. Oil companies were also required to phase in double hulls on oil vessels over a twenty-five-year period by 2015. Double hulls provide an extra container in case of accidents.

Accidents Still Occur Worldwide

In mid-November 2002 the twenty-six-year-old, single-hulled tanker *Prestige* was first damaged, then sank in the Atlantic Ocean off the coast of Spain. The Spanish government estimates the ship spilled approximately 5,000 tons of heavy fuel oil when it was damaged and an additional 12,000 tons as it sank. The ship continued to leak, and the Spanish authorities ordered the leaking vessel to be towed to the open ocean. As of mid-2004 an international team of scientists estimated that the oil spilled from the *Prestige* caused the deaths of nearly 250,000 seabirds. The spill also killed unknown numbers of fish and dolphins, and was responsible for economic damage to the Galician fish and shellfish industries, closing the Spanish coastline to fishing.

CHAPTER 3
NATURAL GAS

Natural gas is an important source of energy in the United States. Methane, ethane, and propane are the primary constituents of natural gas, with methane making up 73 to 95% of the total.

The natural gas industry developed out of the petroleum industry. Wells drilled for oil often produced considerable amounts of natural gas, but early oilmen had no idea what to do with it. Originally considered a waste by-product of oil production, natural gas had no market, nor were transmission lines available to deliver it even if a use had been known. As a result, the gas was burned off, or flared. Pictures of southeast Texas in the early twentieth century show thousands of wooden drilling rigs topped with plumes of gas flaming like burning candles. Even today, flaring sites are sometimes the brightest areas visible in nighttime satellite images, outshining even the largest urban areas.

Nonetheless, researchers soon found ways to use natural gas. In 1925 the first natural gas pipeline, more than 200 miles long, was built from Louisiana to Texas. U.S. demand grew rapidly, especially after World War II. By the 1950s natural gas was providing one-quarter of the nation's energy needs. In the early 2000s natural gas was second only to coal in the share of U.S. energy produced. Crude oil was third. (See Table 1.1 in Chapter 1.) A vast pipeline transmission system connects production facilities in the United States, Canada, and Mexico with natural gas distributors. Figure 3.1 shows the production and consumption figures for natural gas for 2003. Figure 3.2 shows the pattern of natural gas supply and distribution in the United States in 2002.

THE PRODUCTION OF NATURAL GAS

Natural gas is produced from gas and oil wells. There is little delay between production and consumption, except for gas that is placed in storage. Changes in demand are almost immediately reflected by changes in wellhead flows, or supply.

Total U.S. natural gas production in 2003 was 19.1 trillion cubic feet, below the peak levels produced from 1969 to 1975. (See Figure 3.3.) According to the Energy Information Administration (EIA) in its *Annual Energy Review 2003,* Texas, Louisiana, and Oklahoma accounted for 36% of the natural gas produced in the United States in 2003. Although production levels of natural gas are being driven up by increasing demand and rising prices, production continues to be outpaced by consumption. Imported gas makes up the difference between supply and demand.

Natural Gas Wells

In 2003, 366,000 gas wells were in operation in the United States. (See Figure 3.4.) Although the number of producing wells increased steadily after 1960 and more sharply after the mid-1970s, there are slight fluctuations in the number of gas wells in operation year-to-year due to the opening of new wells and the closing of old wells, as well as the effects of weather and economics on well operations.

The average productivity of natural gas wells dropped throughout most of the 1970s and the mid-1980s after hitting peak productivity in 1971; it has remained at a relatively steady low level since then. (See Figure 3.5.)

Offshore Production

Offshore drilling for natural gas accounted for about one-fifth of the total U.S. production in 2003. (See Figure 3.6.) Almost all natural gas produced offshore comes from the Gulf of Mexico and offshore California. U.S. offshore production is expected to increase to meet the nation's growing need for energy, although this type of production could be slowed by environmental restrictions.

Offshore drilling generally occurs on the outer continental shelf, the submerged area offshore with a depth of up to 200 meters (656 feet). Figure 3.7 is a diagram of a continental margin. The continental shelf varies from one coastal area to another. The shelf is relatively narrow along the Pacific coast, wide along much of the Atlantic coast and the Gulf of Alaska, and widest in the Gulf of Mexico. The U.S.

FIGURE 3.1

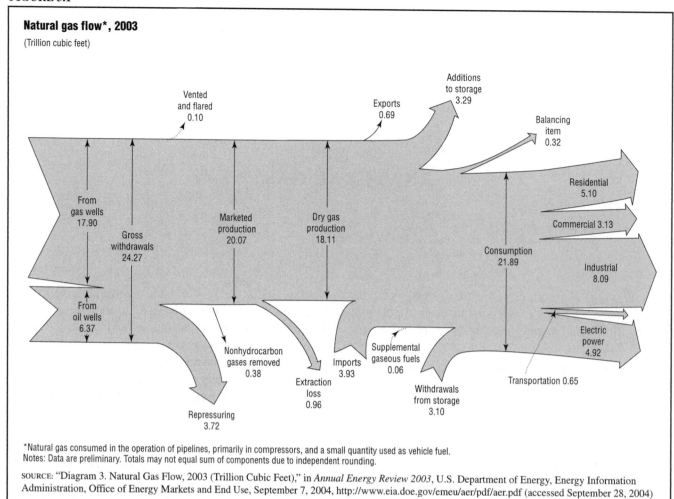

Natural gas flow*, 2003

(Trillion cubic feet)

*Natural gas consumed in the operation of pipelines, primarily in compressors, and a small quantity used as vehicle fuel.
Notes: Data are preliminary. Totals may not equal sum of components due to independent rounding.

SOURCE: "Diagram 3. Natural Gas Flow, 2003 (Trillion Cubic Feet)," in *Annual Energy Review 2003*, U.S. Department of Energy, Energy Information Administration, Office of Energy Markets and End Use, September 7, 2004, http://www.eia.doe.gov/emeu/aer/pdf/aer.pdf (accessed September 28, 2004)

Department of the Interior has leased more than 1.5 billion acres of offshore areas to oil companies for offshore drilling.

The development of offshore oil and gas resources began with the drilling of the Summerland oil field in California in 1896, where about 400 wells were drilled. In the search for oil and gas in offshore areas, the industry has continually improved drilling technology. Today, deepwater petroleum and natural gas exploration occurs from platforms and drill ships, while shallow-water explorations occur from gravel islands and mobile units.

Even though natural gas is transported mostly by pipelines instead of tankers, accidents such as the 1989 *Exxon Valdez* oil spill in Prince William Sound, Alaska, and the *Prestige* oil spill off the coast of Spain in 2002 have focused attention on all types of offshore drilling and tanker transport. Even before the *Exxon Valdez* oil spill, environmentalists were calling for the curtailment of offshore drilling for both oil and gas.

Natural Gas Reserves

Reserves are estimated volumes of gas in known deposits that are believed to be recoverable in the future. Proved reserves are those gas volumes that geological and engineering data show with reasonable certainty to be recoverable. Proved reserves of natural gas amounted to 195.6 trillion cubic feet in 2002. (See Table 3.1.)

Natural gas reserves in North America are generally more abundant than crude oil reserves, although historically they have been difficult to estimate accurately. At one time the U.S. Department of Energy (DOE) estimated that proven supplies of recoverable gas in the United States would last fewer than eight years. But new discoveries and technological improvements have increased the estimated recoverable supply of natural gas to fifty years or more.

The North Slope fields of Alaska are estimated to contain reserves amounting to 35 trillion cubic feet, but at the moment there is no easy way to transport those reserves to the lower forty-eight states. By late 2004 the idea of building a gas pipeline was still being debated. If built, the pipeline could deliver 4.5 billion cubic feet of natural gas per day to the lower forty-eight states—about 10% of the nation's daily gas consumption.

Underground Storage

Because of seasonal, daily, and even hourly changes in gas demand, substantial natural gas storage facilities have

FIGURE 3.2

Natural gas supply and disposition in the United States, 2002

(Trillion cubic feet)

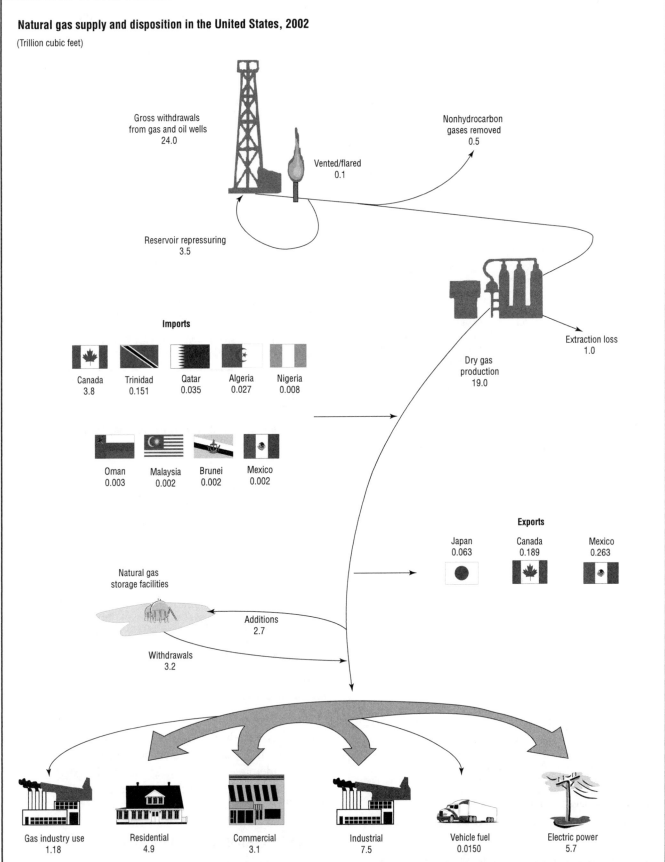

SOURCE: "Figure 2. Natural Gas Supply and Disposition in the United States, 2002 (Trillion Cubic Feet)," in *Natural Gas Annual 2002,* U.S. Department of Energy, Energy Information Administration, Office of Oil and Gas, January 2004, http://www.eia.doe.gov/pub/oil_gas/natural_gas/data_publications/natural_gas_annual/current/pdf/nga02.pdf (accessed November 8, 2004)

FIGURE 3.3

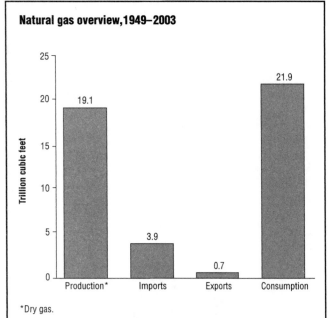

Natural gas overview,1949–2003

*Dry gas.

SOURCE: Adapted from "Figure 6.1. Natural Gas Overview: Overview, 1949–2003," in *Annual Energy Review 2003,* U.S. Department of Energy, Energy Information Administration, Office of Energy Markets and End Use, September 7, 2004, http://www.eia.doe.gov/emeu/aer/pdf/aer.pdf (accessed September 28, 2004)

FIGURE 3.4

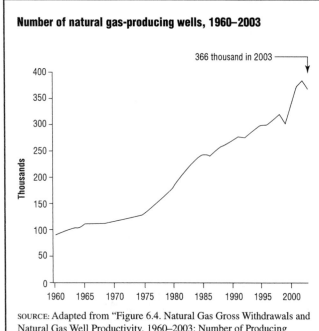

Number of natural gas-producing wells, 1960–2003

SOURCE: Adapted from "Figure 6.4. Natural Gas Gross Withdrawals and Natural Gas Well Productivity, 1960–2003: Number of Producing Wells," in *Annual Energy Review 2003,* U.S. Department of Energy, Energy Information Administration, Office of Energy Markets and End Use, September 7, 2004, http://www.eia.doe.gov/emeu/aer/pdf/aer.pdf (accessed September 28, 2004)

FIGURE 3.5

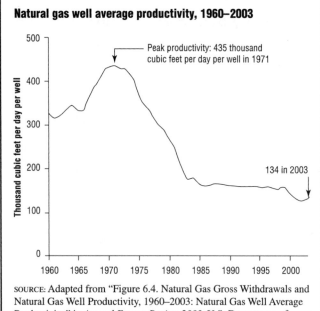

Natural gas well average productivity, 1960–2003

SOURCE: Adapted from "Figure 6.4. Natural Gas Gross Withdrawals and Natural Gas Well Productivity, 1960–2003: Natural Gas Well Average Productivity," in *Annual Energy Review 2003,* U.S. Department of Energy, Energy Information Administration, Office of Energy Markets and End Use, September 7, 2004, http://www.eia.doe.gov/emeu/aer/pdf/aer.pdf (accessed September 28, 2004)

FIGURE 3.6

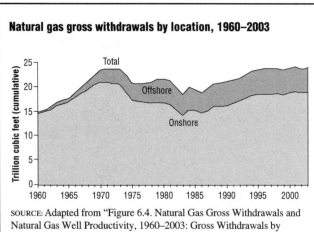

Natural gas gross withdrawals by location, 1960–2003

SOURCE: Adapted from "Figure 6.4. Natural Gas Gross Withdrawals and Natural Gas Well Productivity, 1960–2003: Gross Withdrawals by Location," in *Annual Energy Review 2003,* U.S. Department of Energy, Energy Information Administration, Office of Energy Markets and End Use, September 7, 2004, http://www.eia.doe.gov/emeu/aer/pdf/aer.pdf (accessed September 28, 2004)

been created to meet supply needs. Many of these storage centers are depleted gas reservoirs located near transmission lines and marketing areas. Gas is injected into storage when market needs are lower than the available gas flow,

and gas is withdrawn from storage when supplies from producing fields and/or the capacity of transmission lines are not adequate to meet peak demands. At the end of 2003, gas in underground storage totaled approximately 6.9 trillion cubic feet. (See Figure 3.8.)

TRANSMISSION OF NATURAL GAS

A vast network of natural gas pipelines crisscrosses the United States, connecting every state except Alaska, Hawaii, and Vermont. (Vermont receives its gas directly from Canada and is not connected to the U.S. pipeline.)

FIGURE 3.7

Generalized profile of the continental margin

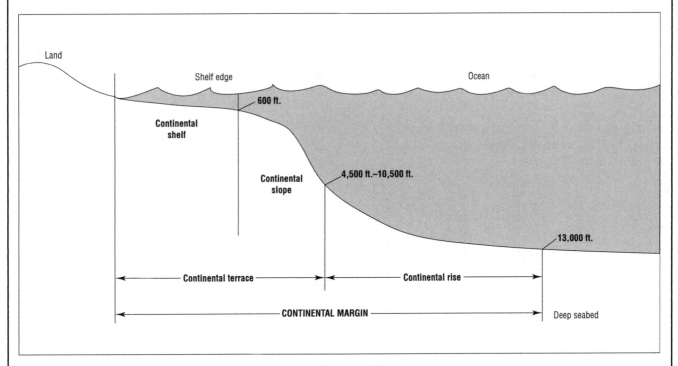

Note: Depths and gradients are approximate.

SOURCE: George Dellagiarino and Keith Meekins, "Figure 1. Profile of the Continental Margin," in *The Resource Evaluation Program: Structure and Mission on the Outer Continental Shelf,* U.S. Department of the Interior, Minerals Management Service, Resource Evaluation Division, 1998, http://www.mms.gov/itd/pubs/1998/98-0028.pdf (accessed November 8, 2004)

The natural gas in this quarter-million-mile system generally flows northeastward, primarily from Texas and Louisiana, the two major gas-producing states, and to a lesser extent from Oklahoma and New Mexico. (See Figure 9.9.) It also flows west to California.

Imports enter the United States via pipeline from Canada in Idaho, Maine, Michigan, Montana, New Hampshire, New York, North Dakota, Washington, and Vermont. It also enters via pipeline in Texas from Mexico. According to *Natural Gas Annual 2002,* published by the EIA in 2004, these pipeline imports made up 94% of total U.S. imports of natural gas in 2002. The remainder is shipped to the United States from Algeria, Australia, Indonesia, Nigeria, Oman, Qatar, Trinidad, and the United Arab Emirates as liquefied natural gas.

The largest users of natural gas in the residential sector in 2002 were, from highest to lowest, Texas, California, Louisiana, New York, and Illinois.

DOMESTIC NATURAL GAS CONSUMPTION

Nationally, natural gas consumption rose from 1949 through 1973, then declined through 1986. Since 1986 natural gas consumption has been generally rising, hitting an all-time high of 23.3 trillion cubic feet in 2000, then declining to 21.9 trillion cubic feet by 2003. (See Table 3.2.) In 2003, 32% of the natural gas was used by industry, 23% by residences, 22% by electric utilities, and 14% by commercial customers; 3% was used as pipeline fuel in the gas transporting process.

Natural gas fills an important part of the country's energy needs. It is an attractive fuel not only because its price is relatively low but also because it is clean and efficient and can help the country meet both its environmental goals and its energy needs.

The residential sector used 5.1 trillion cubic feet of natural gas in 2003. (See Table 3.2.) Energy consumption by residences depends heavily on weather-related home-heating demands. Conservation practices and efficiency of gas appliances such as water heaters and stoves also affect residential consumption patterns. The U.S. Census Bureau reported in its *American Community Survey* that 57% of all residential energy consumers in the United States used gas to heat their homes in 2003.

The use of natural gas in the commercial sector was 3.1 trillion cubic feet in 2003. (See Table 3.2.) Like residential consumption, use in the commercial sector depends heavily on seasonal requirements, as well as the

TABLE 3.1

Crude oil and natural gas field counts, cumulative production, proved reserves, and proved ultimate recovery, 1977–2002

Year	Cumulative number of fields with crude oil and/or natural gas	Cumulative number of fields with crude oil	Crude oil and lease condensate (billion barrels)		Proved ultimate recovery	Cumulative number of fields with natural gas	Natural gas [1] (trillion cubic feet)		Proved ultimate recovery
			Cumulative production	Proved reserves			Cumulative production	Proved reserves	
1977	31,360	27,835	121.4	33.6	155.0	23,883	558.3	209.5	767.8
1978	32,430	28,683	124.6	33.1	157.6	24,786	578.4	210.1	788.5
1979	33,644	29,671	127.7	31.2	158.9	25,823	599.1	208.3	807.4
1980	34,999	30,766	130.8	31.3	162.2	26,919	619.4	206.3	825.6
1981	36,621	32,111	133.9	31.0	165.0	28,213	639.4	209.4	848.9
1982	38,123	33,375	137.1	29.5	166.6	29,375	658.1	209.3	867.4
1983	39,489	34,495	140.3	29.3	169.6	30,419	675.1	209.0	884.1
1984	41,038	35,784	143.5	30.0	173.5	31,595	693.5	206.0	899.5
1985	42,317	36,849	146.8	29.9	176.7	32,595	710.9	202.2	913.1
1986	43,076	37,464	150.0	28.3	178.3	33,151	727.8	201.1	928.9
1987	43,742	37,982	153.0	28.7	181.7	33,657	745.4	196.4	941.8
1988	44,414	38,506	156.0	28.2	184.2	34,196	763.4	177.0	940.4
1989	44,883	38,858	158.8	27.9	186.7	34,579	781.7	175.4	957.1
1990	45,385	39,244	161.5	27.6	189.0	34,975	800.4	177.6	978.0
1991	45,776	39,558	164.2	25.9	190.1	35,254	819.1	175.3	994.4
1992	46,149	39,843	166.8	25.0	191.8	35,539	838.0	173.3	1,011.3
1993	46,513	40,124	169.3	24.1	193.4	35,798	857.2	170.5	1,027.7
1994	46,922	40,417	171.7	23.6	195.3	36,142	877.1	171.9	1,049.1
1995	47,296	40,694	174.1	23.5	197.7	36,433	896.9	173.5	1,070.4
1996	47,557	40,875	176.5	23.3	199.8	36,612	917.0	175.1	1,092.1
1997	47,854	40,977	178.9	23.9	202.8	36,830	937.1	175.7	1,112.8
1998	[2]47,664	[2]35,143	181.2	22.4	203.5	[2]32,458	957.0	172.4	1,129.4
1999	NA	NA	183.3	23.2	206.5	NA	976.8	176.2	1,153.0
2000	NA	NA	185.4	23.5	208.9	NA	997.0	186.5	1,183.5
2001	NA	NA	187.5	R23.9	R211.4	NA	1,016.7	183.5	1,200.2
2002	NA	NA	189.6	24.0	213.6	NA	1,036.9	195.6	1,232.5

[1]Wet, after separation of lease condensate.
[2]There is a discontinuity in this time series between 1997 and 1998 due to the absence of updates for a subset of the data used in the past.
R=Revised.
NA=Not available.
Notes: Data are at end of year. See "Proved Reserves, Crude Oil," "Proved Reserves, Lease Condensate," "Proved Reserves, Natural Gas," and "Proved Reserves, Natural Gas Liquids" in Glossary.
Web Pages: See http://www.eia.doe.gov/oil_gas/petroleum/info_glance/petroleum.html and http://www.eia.doe.gov/oil_gas/natural_gas/info_glance/natural_gas.html for related information.

SOURCE: "Table 4.2. Crude Oil and Natural Gas Field Counts, Cumulative Production, Proved Reserves, and Proved Ultimate Recovery, 1977–2002," in *Annual Energy Review 2003*, U.S. Department of Energy, Energy Information Administration, Office of Energy Markets and End Use, September 7, 2004, http://www.eia.doe.gov/emeu/aer/pdf/aer.pdf (accessed September 28, 2004)

number of users and conservation measures taken by commercial establishments.

The industrial sector has historically been the largest consumer of natural gas. Consumption in this sector in 2003 was 8.1 trillion cubic feet, down from 8.6 trillion cubic feet in 2002. The all-time high of 10.2 trillion cubic feet occurred in 1973. (See Table 3.2.) After 1973 natural gas consumption declined through 1986, steadily increased through 2000, and then declined somewhat through 2003. Substitution of natural gas for petroleum for some industrial purposes caused much of the increase in natural gas consumption from 1986 through 2000.

NATURAL GAS PRICES

Natural gas prices can vary because of differing federal and state rate structures. Region also plays a role—for example, prices are lower in major natural gas-producing areas where transmission costs are lower.

From the mid-twentieth century through the early 1970s, natural gas prices were relatively stable. (See Figure 3.10.) Thereafter, deregulation and industry restructuring brought about a period of sharply rising prices, with wellhead prices (the value of natural gas at the mouth of the well) reaching a high in 1983, declining until 1991, and then generally increasing through 2003. The average price of all categories of natural gas at the wellhead in 2003 was $4.71 per 1,000 cubic feet, sharply rising from $2.84 in 2002.

At the retail price level in real dollars, residential customers paid $8.99 per thousand cubic feet of natural gas in 2003 and $7.61 in 2002. (See Table 3.3.) Commercial consumers paid $7.82 per thousand cubic feet in 2003, while industrial consumers paid $5.47 per thousand cubic feet.

Much of the variation in natural gas prices through the years can be attributed to changes in the natural gas industry. The passage of the Natural Gas Policy Act of

FIGURE 3.8

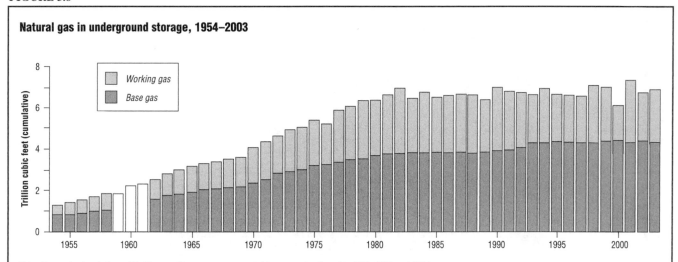

Natural gas in underground storage, 1954–2003

Note: Storage is at end of year. Working- and base-gas component data were not collected in 1959, 1960, and 1961.

SOURCE: Adapted from "Figure 6.6. Natural Gas in Underground Storage, 1954–2003: Total," in *Annual Energy Review 2003,* U.S. Department of Energy, Energy Information Administration, Office of Energy Markets and End Use, September 7, 2004, http://www.eia.doe.gov/emeu/aer/pdf/aer.pdf (accessed September 28, 2004)

FIGURE 3.9

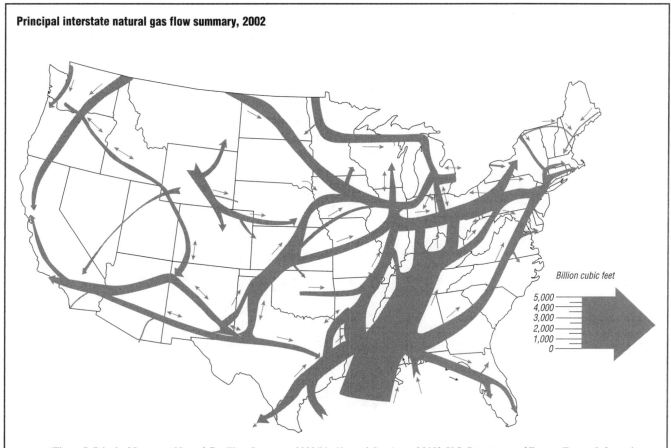

Principal interstate natural gas flow summary, 2002

SOURCE: "Figure 7. Principal Interstate Natural Gas Flow Summary, 2002," in *Natural Gas Annual 2002,* U.S. Department of Energy, Energy Information Administration, Office of Oil and Gas, January 2004, http://www.eia.doe.gov/pub/oil_gas/natural_gas/data_publications/natural_gas_annual/current/pdf/nga02.pdf (accessed November 8, 2004)

1978 (NGPA; PL 95–621) triggered a dramatic transformation in the natural gas industry. The NGPA allowed gas prices at the wellhead to rise gradually. (See Figure 3.10.)

On January 1, 1985, new gas prices were deregulated, and additional volumes of onshore production were deregulated on July 1, 1987. In 1988 President Ronald Reagan

TABLE 3.2

Natural gas consumption by sector, selected years, 1949–2003

(Billion cubic feet)

		Commercial			End-use sectors — Industrial					Transportation				Electric power sector[1]			
						Other industrial											
Year	Resi-dential	CHP[2]	Other[3]	Total	Lease and plant fuel	CHP[4]	Non-CHP[5]	Total	Total	Pipeline fuel[6]	Vehicle fuel	Total	Total	Electricity only	CHP	Total	Total
1949	993	[7]	348	348	835	[8]	2,245	2,245	3,081	NA	NA	NA	4,421	550	NA	550	4,971
1950	1,198	[7]	388	388	928	[8]	2,498	2,498	3,426	126	NA	126	5,138	629	NA	629	5,767
1955	2,124	[7]	629	629	1,131	[8]	3,411	3,411	4,542	245	NA	245	7,540	1,153	NA	1,153	8,694
1960	3,103	[7]	1,020	1,020	1,237	[8]	4,535	4,535	5,771	347	NA	347	10,242	1,725	NA	1,725	11,967
1965	3,903	[7]	1,444	1,444	1,156	[8]	5,955	5,955	7,112	501	NA	501	12,959	2,321	NA	2,321	15,280
1970	4,837	[7]	2,399	2,399	1,399	[8]	7,851	7,851	9,249	722	NA	722	17,208	3,932	NA	3,932	21,139
1971	4,972	[7]	2,509	2,509	1,414	[8]	8,181	8,181	9,594	743	NA	743	17,817	3,976	NA	3,976	21,793
1972	5,126	[7]	2,608	2,608	1,456	[8]	8,169	8,169	9,624	766	NA	766	18,125	3,977	NA	3,977	22,101
1973	4,879	[7]	2,597	2,597	1,496	[8]	8,689	8,689	10,185	728	NA	728	18,389	3,660	NA	3,660	22,049
1974	4,786	[7]	2,556	2,556	1,477	[8]	8,292	8,292	9,769	669	NA	669	17,780	3,443	NA	3,443	21,223
1975	4,924	[7]	2,508	2,508	1,396	[8]	6,968	6,968	8,365	583	NA	583	16,380	3,158	NA	3,158	19,538
1976	5,051	[7]	2,668	2,668	1,634	[8]	6,964	6,964	8,598	548	NA	548	16,866	3,081	NA	3,081	19,946
1977	4,821	[7]	2,501	2,501	1,659	[8]	6,815	6,815	8,474	533	NA	533	16,329	3,191	NA	3,191	19,521
1978	4,903	[7]	2,601	2,601	1,648	[8]	6,757	6,757	8,405	530	NA	530	16,439	3,188	NA	3,188	19,627
1979	4,965	[7]	2,786	2,786	1,499	[8]	6,899	6,899	8,398	601	NA	601	16,750	3,491	NA	3,491	20,241
1980	4,752	[7]	2,611	2,611	1,026	[8]	7,172	7,172	8,198	635	NA	635	16,196	3,682	NA	3,682	19,877
1981	4,546	[7]	2,520	2,520	928	[8]	7,128	7,128	8,055	642	NA	642	15,764	3,640	NA	3,640	19,404
1982	4,633	[7]	2,606	2,606	1,109	[8]	5,831	5,831	6,941	596	NA	596	14,776	3,226	NA	3,226	18,001
1983	4,381	[7]	2,433	2,433	978	[8]	5,643	5,643	6,621	490	NA	490	13,924	2,911	NA	2,911	16,835
1984	4,555	[7]	2,524	2,524	1,077	[8]	6,154	6,154	7,231	529	NA	529	14,839	3,111	NA	3,111	17,951
1985	4,433	[7]	2,432	2,432	966	[8]	5,901	5,901	6,867	504	NA	504	14,237	3,044	NA	3,044	17,281
1986	4,314	[7]	2,318	2,318	923	[8]	5,579	5,579	6,502	485	NA	485	13,619	2,602	NA	2,602	16,221
1987	4,315	[7]	2,430	2,430	1,149	[8]	5,953	5,953	7,103	519	NA	519	14,367	2,844	NA	2,844	17,211
1988	4,630	[7]	2,670	2,670	1,096	[8]	6,383	6,383	7,479	614	NA	614	15,394	2,636	NA	2,636	18,030
1989	4,781	30	2,688	2,718	1,070	914	[9]5,903	[9]6,816	7,886	629	NA	629	16,014	[9]2,791	[9]315	[9]3,105	[9]19,119
1990	4,391	46	2,576	2,623	1,236	1,055	[9]5,963	[9]7,018	8,255	660	(s)	660	15,929	[9]2,794	[9]451	[9]3,245	[9]19,174
1991	4,556	52	2,676	2,729	1,129	1,061	[9]6,170	[9]7,231	8,360	601	(s)	602	16,246	[9]2,822	[9]494	[9]3,316	[9]19,562
1992	4,690	62	2,740	2,803	1,171	R1,107	[9]R6,420	[9]7,527	8,698	588	2	590	16,780	[9]2,829	[9]619	[9]3,448	[9]20,228
1993	4,956	65	2,796	2,862	1,172	R1,124	R6,576	7,700	8,872	624	3	627	17,317	2,755	718	3,473	20,790
1994	4,848	72	2,823	2,895	1,124	R1,176	R6,613	7,790	8,913	685	3	689	17,345	3,065	838	3,903	21,247
1995	4,850	78	2,953	3,031	1,220	R1,258	R6,906	8,164	9,384	700	5	705	17,970	3,288	949	4,237	22,207
1996	5,241	82	3,076	3,158	1,250	1,289	7,146	8,435	9,685	711	6	718	18,802	2,824	983	3,807	22,609
1997	4,984	87	3,128	3,215	1,203	1,282	7,229	8,511	9,714	751	8	760	18,673	3,039	1,026	4,065	22,737
1998	4,520	87	2,912	2,999	1,173	1,355	6,965	8,320	9,493	635	9	645	17,658	3,544	1,044	4,588	22,246
1999	4,726	84	2,961	3,045	1,079	1,401	6,678	8,079	9,158	645	[10]12	657	17,586	3,729	1,090	4,820	22,405
2000	4,996	85	R3,098	R3,182	1,151	1,386	6,757	8,142	9,293	642	[10]13	655	R18,127	4,093	1,114	5,206	R23,333
2001	R4,771	79	R2,944	R3,023	R1,119	R1,310	R6,035	R7,344	R8,463	R625	[10]15	R640	R16,896	4,164	R1,178	R5,342	R22,239
2002	R4,890	R74	3,029	R3,103	R1,114	R1,240	R6,316	R7,557	R8,671	R667	[10]15	R682	R17,346	R4,258	R1,413	R5,672	R23,018
2003P	5,101	71	3,058	3,129	1,123	1,138	5,829	6,967	8,090	635	[10]15	650	16,970	3,611	1,313	4,924	21,894

TABLE 3.2

Natural gas consumption by sector, selected years, 1949–2003 [CONTINUED]

(Billion cubic feet)

[1]Electricity-only and combined-heat-and-power (CHP) plants within the NAICS (North American Industry Classification System) 22 category whose primary business is to sell electricity, or electricity and heat, to the public. Through 1988, data are for electric utilities only; beginning in 1989, data are for electric utilities and independent power producers. Electric utility CHP plants are included in "Electricity Only."

[2]Commercial combined-heat-and-power and a small number of commercial electricity-only plants.

[3]All commercial sector fuel use other than that in "Commercial CHP."

[4]Industrial combined-heat-and-power (CHP) and a small number of industrial electricity-only plants.

[5]All industrial sector fuel use other than that in "Lease and plant fuel" and "Industrial CHP."

[6]Natural gas consumed in the operation of pipelines, primarily in compressors.

[7]Included in "Commercial other."

[8]Included in "Industrial non–CHP."

[9]For 1989–1992, a small amount of consumption at independent power producers may be counted in other "Other industrial" and "Electric power sector."

[10]For 1999 forward, vehicle fuel data do not reflect revised data. These revisions, in million cubic feet, are: 1999—10,313; 2000—11,365; 2001—13,646; 2002—15,657; and 2003—18,339.

R = Revised.

P = Preliminary.

NA = Not available.

(s) = Less than 0.5 billion cubic feet.

Notes: Data are for natural gas, plus a small amount of supplemental gaseous fuels that cannot be identified separately. Beginning with 1965, all volumes are shown on a pressure base of 14.73 p.s.i.a. at 60° F For prior years, the pressure base was 14.65 p.s.i.a. at 60° F. Totals may not equal sum of components due to independent rounding. Web Pages: For data not shown for 1951–1969, see http://www.eia.doe.gov/emeu/aer/natgas.html. For related information, see http://www.eia.doe.gov/oil_gas/natural_gas/info_glance/natural_gas.html.

SOURCE: "Table 6.5. Natural Gas Consumption by Sector, Selected Years, 1949– 2003 (Billion Cubic Feet)," in *Annual Energy Review 2003*, U.S. Department of Energy, Office of Energy Markets and End Use, Energy Information Administration, September 7, 2004, http://www.eia.doe.gov/emeu/aer/pdf/aer.pdf (accessed September 28, 2004)

FIGURE 3.10

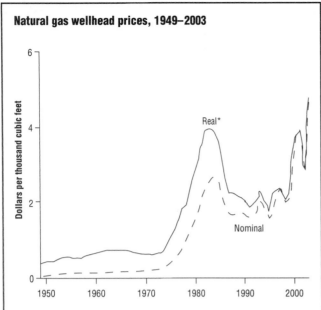

Natural gas wellhead prices, 1949–2003

*In chained (2000) dollars, calculated by using gross domestic product implicit price deflators.

SOURCE: Adapted from "Figure 6.7. Natural Gas Wellhead, City Gate, and Imports Prices: Wellhead, 1949–2003," in *Annual Energy Review 2003*, U.S. Department of Energy, Energy Information Administration, Office of Energy Markets and End Use, September 7, 2004, http://www.eia.doe.gov/emeu/aer/pdf/aer.pdf (accessed September 28, 2004)

signed legislation removing all remaining natural gas wellhead price controls by 1993.

The NGPA allowed prices to go up, but it also opened the market to the forces of supply and demand. Now that prices are deregulated and the industry is no longer constrained by federal controls, the natural gas industry has become more sensitive to market signals and is able to respond more quickly to changes in economic conditions.

NATURAL GAS IMPORTS AND EXPORTS

U.S. natural gas trading was limited to the neighboring countries of Mexico and Canada until shipping natural gas in liquefied form became a feasible alternative to pipelines. In 1969 the first shipments of liquefied natural gas (LNG) were sent from Alaska to Japan, and U.S. imports of LNG from Algeria began the following year.

In 2003 U.S. net imports of natural gas by all routes totaled 3.2 trillion cubic feet, 14.8% of domestic consumption. Natural gas imports have been increasing significantly since 1986. Historically, Canada has been by far the major supplier of U.S. natural gas imports, accounting for 87% of the natural gas imported in 2003 (see Figure 3.11.)

The EIA reported in *Annual Energy Review 2003* (published in 2004) that the United States exported 692 billion cubic feet of natural gas in 2003. Of these exports, Mexico bought the largest amount (333 billion cubic

feet), while Canada purchased 294 billion cubic feet, and Japan bought 64 billion cubic feet.

INTERNATIONAL NATURAL GAS USAGE

World Production

World production of dry natural gas totaled an all-time high of 92.2 trillion cubic feet in 2002. (See Table 3.4.) Russia and the United States were the largest producers, with Russia accounting for 21 trillion cubic feet and the United States producing 19 trillion cubic feet.

World Consumption

World consumption of natural gas has increased steadily since 1980, from 52.9 trillion cubic feet to 92.1 trillion cubic feet in 2002. The United States consumed the largest amount of natural gas in 2002, followed by Russia. (See Figure 3.12.) Combined, they accounted for 41% of world consumption.

FUTURE TRENDS IN THE GAS INDUSTRY

In its 2004 publication *Annual Energy Outlook 2004*, the EIA projects energy supply, demand, and prices through 2025, predicting that natural gas production will increase steadily, as will consumption, pipeline expansion, and imports. Natural gas prices for residential customers are projected to rise by 9% from 2002 to 2025.

Domestic Production

Total domestic natural gas production is projected to increase from 2002 levels of 19.1 trillion cubic feet to 24 trillion cubic feet in 2025, with an annual growth rate of about 1%. Domestic production will be boosted by increases from onshore sources in the lower forty-eight states, slight offshore sources in the Gulf of Mexico, and significant increases resulting from the advent of North Slope gas pipeline operations in Alaska around 2018.

Figure 3.13 shows projected figures for different types of natural gas production. The letters "NA" and "AD" in Figure 3.13 stand for nonassociated (NA) and associated-dissolved (AD) natural gas. These terms refer to natural gas that is found in conjunction with crude oil (associated) or not in conjunction with crude oil (nonassociated). Associated-dissolved natural gas is found in a dissolved state with the oil, like oxygen dissolved in aquarium water. Unconventional sources of natural gas are those from which it is more difficult and less economically sound to extract natural gas because the technology to reach it has not been developed fully or is too expensive.

Domestic Consumption

The *Annual Energy Outlook 2004* projects that consumption of natural gas will outpace production from 2002 to 2025. Total domestic natural gas consumption is projected to increase from 2002 levels of 22.8 trillion cubic

TABLE 3.3

Natural gas prices by sector, 1967–2003

(Dollars per thousand cubic feet)

Year	Residential Prices Nominal	Residential Prices Real	Residential Percentage of sector	Commercial Prices Nominal	Commercial Prices Real	Commercial Percentage of sector	Industrial Prices Nominal	Industrial Prices Real	Industrial Percentage of sector	Vehicle fuel Prices Nominal	Vehicle fuel Prices Real	Electric power Prices Nominal	Electric power Prices Real	Electric power Percentage of sector
1967	1.04	R4.35	NA	0.74	R3.10	NA	0.34	R1.42	NA	NA	NA	0.28	R1.17	NA
1968	1.04	R4.17	NA	0.73	R2.93	NA	0.34	R1.36	NA	NA	NA	0.22	R0.88	NA
1969	1.05	R4.02	NA	0.74	R2.83	NA	0.35	R1.34	NA	NA	NA	0.27	R1.03	NA
1970	1.09	R3.96	NA	0.77	R2.80	NA	0.37	R1.34	NA	NA	NA	0.29	R1.05	NA
1971	1.15	R3.98	NA	0.82	R2.84	NA	0.41	R1.42	NA	NA	NA	0.32	R1.11	NA
1972	1.21	R4.01	NA	0.88	R2.92	NA	0.45	R1.49	NA	NA	NA	0.34	R1.13	NA
1973	1.29	R4.05	NA	0.94	R2.95	NA	0.50	R1.57	NA	NA	NA	0.38	R1.19	92.1
1974	1.43	R4.12	NA	1.07	R3.08	NA	0.67	R1.93	NA	NA	NA	0.51	R1.47	92.7
1975	1.71	R4.50	NA	1.35	R3.55	NA	0.96	R2.53	NA	NA	NA	0.77	R2.03	96.1
1976	1.98	R4.93	NA	1.64	R4.08	NA	1.24	R3.08	NA	NA	NA	1.06	R2.64	96.2
1977	2.35	R5.50	NA	2.04	R4.77	NA	1.50	R3.51	NA	NA	NA	1.32	R3.09	97.1
1978	2.56	R5.59	NA	2.23	R4.87	NA	1.70	R3.72	NA	NA	NA	1.48	R3.23	98.0
1979	2.98	R6.01	NA	2.73	R5.51	NA	1.99	R4.02	NA	NA	NA	1.81	R3.65	96.1
1980	3.68	R6.81	NA	3.39	R6.27	NA	2.56	R4.74	NA	NA	NA	2.27	R4.20	96.9
1981	4.29	R7.26	NA	4.00	R6.77	NA	3.14	R5.31	NA	NA	NA	2.89	R4.89	97.6
1982	5.17	R8.24	NA	4.82	R7.68	NA	3.87	R6.17	85.1	NA	NA	3.48	R5.55	92.6
1983	6.06	R9.29	NA	5.59	R8.57	NA	4.18	R6.41	80.7	NA	NA	3.58	R5.49	93.9
1984	6.12	R9.05	NA	5.55	R8.20	NA	4.22	R6.24	74.7	NA	NA	3.70	R5.47	94.4
1985	6.12	R8.78	NA	5.50	R7.89	NA	3.95	R5.67	68.8	NA	NA	3.55	R5.09	94.0
1986	5.83	R8.18	NA	5.08	R7.13	NA	3.23	R4.53	59.8	NA	NA	2.43	R3.41	91.7
1987	5.54	R7.57	NA	4.77	R6.52	93.1	2.94	R4.02	47.4	NA	NA	2.32	R3.17	91.6
1988	5.47	R7.23	NA	4.63	R6.12	90.7	2.95	R3.90	42.6	NA	NA	2.33	R3.08	89.6
1989	5.64	R7.18	99.9	4.74	R6.03	89.1	2.96	R3.77	36.9	NA	NA	2.43	R3.09	88.6
1990	5.80	R7.11	99.3	4.83	R5.92	86.6	2.93	R3.59	35.2	3.39	R4.15	2.38	R2.92	89.2
1991	5.82	R6.89	99.2	4.81	R5.70	85.1	2.69	R3.19	32.7	3.96	R4.69	2.18	R2.58	93.2
1992	5.89	R6.82	99.1	4.88	R5.65	83.2	2.84	R3.29	30.3	4.05	R4.69	2.36	R2.73	93.2
1993	6.16	R6.97	99.1	5.22	R5.91	83.9	3.07	R3.47	29.7	4.27	R4.83	2.61	R2.95	93.4
1994	6.41	R7.10	99.1	5.44	R6.03	79.3	3.05	R3.38	25.5	4.11	R4.55	2.28	R2.53	93.5
1995	6.06	R6.58	99.1	5.05	R5.48	76.7	2.71	R2.94	24.5	3.98	R4.32	2.02	R2.19	92.0
1996	6.34	R6.76	99.1	5.40	R5.75	77.6	3.42	R3.64	19.4	4.34	R4.62	2.69	R2.87	92.2
1997	6.94	R7.27	98.8	5.80	R6.08	70.8	3.59	R3.76	18.1	4.44	R4.65	2.78	R2.91	91.0
1998	6.82	R7.07	97.7	5.48	R5.68	67.0	3.14	R3.25	16.1	4.59	R4.76	2.40	R2.49	82.5
1999	6.69	R6.84	95.2	5.33	R5.45	66.1	3.12	R3.19	18.8	4.34	R4.43	2.62	R2.68	75.3
2000	7.76	R7.76	92.6	6.59	R6.59	R63.9	4.45	R4.45	19.8	5.54	R5.54	4.38	R4.38	64.3
2001	R9.63	9.41	R92.4	8.43	8.23	R66.0	R5.24	R5.12	R20.8	6.60	R6.45	4.61	R4.50	41.9
2002	R7.91	R7.61	R91.4	R6.64	R6.39	R78.4	R4.02	R3.87	R22.5	R4.74	R4.56	5,R3.68	5,R3.54	5 81.1
2003	P9.50	P8.99	E92.1	P8.26	P7.82	P77.2	P5.78	P5.47	P22.2	NA	NA	P5.57	P5.27	P83.6

TABLE 3.3

Natural gas prices by sector, 1967–2003 [CONTINUED]

(Dollars per thousand cubic feet)

[1]Residential, commercial, and industrial prices do not include the price of natural gas delivered to consumers on behalf of third parties.
[2]Commercial sector, including commercial combined-heat-and-power (CHP) and commercial electricity-only plants.
[3]Industrial sector, including industrial combined-heat-and-power (CHP) and industrial electricity-only plants.
[4]Much of the natural gas delivered for vehicle fuel represents deliveries to fueling stations that are used primarily or exclusively by respondents' fleet vehicles. Thus, the prices are often those associated with the operation of fleet vehicles.
[5]Electricity-only and combined-heat-and-power (CHP) plants within the NAICS (North American Industry Classification System) 22 category whose primary business is to sell electricity, or electricity and heat, to the public. Through 2001, data are for electric utilities only; beginning in 2002, data are for electric utilities and independent power producers.
[6]In chained (2000) dollars, calculated by using gross domestic product implicit price deflators.
R=Revised.
P=Preliminary.
E=Estimate.
NA=Not available.
Notes: Prices are for natural gas, plus a small amount of supplemental gaseous fuels that cannot be identified separately. The average for each end-use sector is calculated by dividing the total value of the natural gas consumed by each sector by the total quantity consumed. Prices are intended to include all taxes.
Web Page: See http://www.eia.doe.gov/oil_gas/natural_gas/info_glance/natural_gas.html for related information.

SOURCE: "Table 6.8. Natural Gas Prices by Sector, 1967–2003 (Dollars per Thousand Cubic Feet)," in *Annual Energy Review 2003*, U.S. Department of Energy, Energy Information Administration, Office of Energy Markets and End Use, September 7, 2004, http://www.eia.doe.gov/emeu/aer/pdf/aer.pdf (accessed September 28, 2004)

FIGURE 3.11

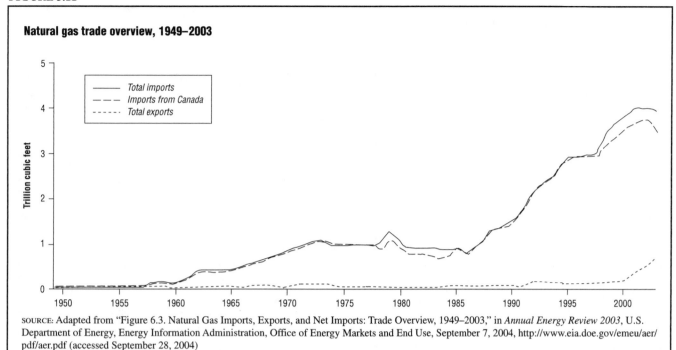

Natural gas trade overview, 1949–2003

SOURCE: Adapted from "Figure 6.3. Natural Gas Imports, Exports, and Net Imports: Trade Overview, 1949–2003," in *Annual Energy Review 2003*, U.S. Department of Energy, Energy Information Administration, Office of Energy Markets and End Use, September 7, 2004, http://www.eia.doe.gov/emeu/aer/pdf/aer.pdf (accessed September 28, 2004)

feet to 31.4 trillion cubic feet in 2025, with an annual growth rate of 1.4%. Demand for natural gas by industrial consumers is expected to account for approximately 33% of total consumption in 2025, electricity producers 30%, residential 19%, and commercial 13%. (See Figure 3.14.) Natural gas will be called upon to replace the nation's aging nuclear electricity plants because of its efficiency and low emissions. To meet growing demand, natural gas pipeline capacity will have to be expanded, particularly along the corridors that move Canadian supplies to the Pacific Coast and the corridors that move Pacific Coast Canadian imports and Gulf of Mexico offshore natural gas to the East. A new pipeline from the North Slope in Alaska is expected to be constructed well before 2025.

Imports and Exports

Net imports of natural gas are projected to increase to meet demand, from 3.5 trillion cubic feet in 2002 to 7.2 trillion cubic feet in 2025. Most of these imports will come from Canada. (See Figure 3.15.) The rest will be shipped to the United States from Algeria, Australia, Indonesia, Nigeria, Oman, Qatar, Trinidad, and the United Arab Emirates in the form of liquefied natural gas. Natural gas exports to Mexico are expected to continue and to peak in 2006 (shown as an import minus value in Figure 3.15.) After that time exports to Mexico are expected to decline; Mexico's own natural gas infrastructure should be developed by 2015, and it is expected to begin meeting its own natural gas needs.

TABLE 3.4

World dry natural gas production, 1993–2002

(Trillion cubic feet)

Region and country	1993	1994	1995	1996	1997	1998	1999	2000	2001	2002ᴾ
North, Central, and South America	**26.26**	**27.50**	**27.74**	**28.39**	**28.75**	**29.39**	**29.53**	**30.39**	ᴿ**31.17**	**30.65**
Argentina	0.76	0.79	0.88	0.94	0.97	1.04	1.22	1.32	1.31	1.28
Canada	4.91	5.27	5.60	5.71	5.76	5.98	6.26	6.47	6.60	6.63
Mexico	0.95	0.97	0.96	1.06	1.17	1.27	1.29	1.31	1.30	1.33
United States	18.10	18.82	18.60	18.85	18.90	19.02	18.83	19.18	ᴿ19.62	18.96
Venezuela	0.82	0.88	0.89	0.96	0.99	1.11	0.95	0.96	1.12	1.05
Other	0.73	0.78	0.81	0.86	0.96	0.96	0.98	1.15	ᴿ1.22	1.39
Western Europe	**8.33**	**8.44**	**8.80**	**10.09**	**9.71**	**9.64**	**9.92**	**10.19**	ᴿ**10.27**	**10.55**
Germany	0.68	0.70	0.74	0.80	0.79	0.77	0.82	0.78	ᴿ0.79	0.79
Italy	0.69	0.73	0.72	0.71	0.68	0.67	0.62	0.59	ᴿ0.54	0.51
Netherlands	3.11	2.95	2.98	3.37	2.99	2.84	2.67	2.56	2.75	2.66
Norway	0.97	1.04	1.08	1.45	1.62	1.63	1.76	1.87	ᴿ1.95	2.41
United Kingdom	2.31	2.47	2.67	3.18	3.03	3.14	3.49	3.83	ᴿ3.69	3.61
Other	0.57	0.55	0.61	0.59	0.60	0.58	0.57	0.57	0.57	0.57
Eastern Europe and Former U.S.S.R.	**27.99**	**26.47**	**25.93**	**26.28**	**24.85**	**25.17**	**25.41**	**26.22**	**26.48**	**27.05**
Romania	0.75	0.69	0.68	0.63	0.61	0.52	0.50	0.48	0.51	0.47
Russia	21.81	21.45	21.01	21.23	20.17	20.87	20.83	20.63	20.51	21.03
Turkmenistan	2.29	1.26	1.14	1.31	0.90	0.47	0.79	1.64	1.70	1.89
Ukraine	0.68	0.64	0.62	0.64	0.64	0.64	0.63	0.64	0.64	0.65
Uzbekistan	1.59	1.67	1.70	1.70	1.74	1.94	1.96	1.99	2.23	2.04
Other	0.87	0.76	0.79	0.76	0.79	0.74	0.70	0.84	0.89	0.97
Middle East and Africa	**7.24**	**7.41**	**7.99**	**8.76**	**9.74**	**10.30**	**10.95**	**12.01**	ᴿ**12.61**	**13.41**
Algeria	1.90	1.81	2.05	2.19	2.43	2.60	2.88	2.94	ᴿ2.79	2.80
Egypt	0.40	0.42	0.44	0.47	0.48	0.49	0.52	0.65	ᴿ0.87	0.94
Iran	0.96	1.12	1.25	1.42	1.66	1.77	2.04	2.13	ᴿ2.33	2.65
Qatar	0.48	0.48	0.48	0.48	0.61	0.69	0.78	1.03	ᴿ0.95	1.04
Saudi Arabia	1.27	1.33	1.34	1.46	1.60	1.65	1.63	1.76	1.90	2.00
United Arab Emirates	0.94	0.91	1.11	1.19	1.28	1.31	1.34	1.36	ᴿ1.39	1.53
Other	1.30	1.34	1.33	1.53	1.67	1.79	1.76	2.15	ᴿ2.39	2.45
Asia and Oceania	**6.55**	**7.11**	**7.50**	**8.13**	**8.47**	**8.55**	**9.10**	**9.48**	ᴿ**9.92**	**10.55**
Australia	0.86	0.93	1.03	1.06	1.06	1.10	1.10	1.16	ᴿ1.19	1.26
China	0.56	0.59	0.60	0.67	0.75	0.78	0.85	0.96	1.07	1.15
India	0.53	0.59	0.63	0.70	0.72	0.76	0.75	0.79	ᴿ0.85	0.88
Indonesia	1.97	2.21	2.24	2.35	2.37	2.27	2.51	2.36	ᴿ2.34	2.48
Malaysia	0.88	0.92	1.02	1.23	1.36	1.37	1.42	1.50	ᴿ1.66	1.71
Pakistan	0.58	0.63	0.65	0.70	0.70	0.71	0.78	0.86	ᴿ0.77	0.81
Other	1.16	1.23	1.33	1.42	1.52	1.56	1.69	1.86	2.04	2.25
World	**76.36**	**76.93**	**77.96**	**81.65**	**81.52**	**83.03**	**84.91**	**88.28**	ᴿ**90.45**	**92.20**

R=Revised.
P=Preliminary.
Note: Totals may not equal sum of components due to independent rounding.
Web Page: For related information, see http://www.eia.doe.gov/international.

SOURCE: "Table 11.11. World Dry Natural Gas Production, 1993–2002 (Trillion Cubic Feet)," in *Annual Energy Review 2003*, U.S. Department of Energy, Energy Information Administration, Office of Energy Markets and End Use, September 7, 2004, http://www.eia.doe.gov/emeu/aer/pdf/aer.pdf (accessed September 28, 2004)

FIGURE 3.12

Selected consumers of dry natural gas, by country, 2002

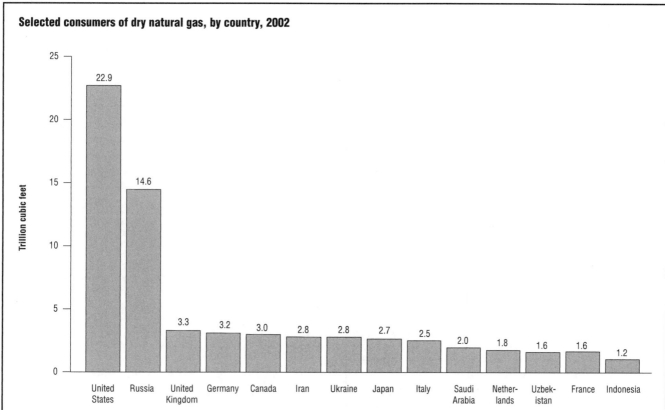

SOURCE: Adapted from "Figure 11.12. World Dry Natural Gas Consumption: Selected Consuming Countries, 2002," in *Annual Energy Review 2003,* U.S. Department of Energy, Energy Information Administration, Office of Energy Markets and End Use, September 7, 2004, http://www.eia.doe.gov/emeu/aer/pdf/aer.pdf (accessed September 28, 2004)

FIGURE 3.13

Natural gas production by source, 1990–2025

SOURCE: "Figure 87. Natural Gas Production by Source, 1990–2025 (Trillion Cubic Feet)," in *Annual Energy Outlook 2004,* U.S. Department of Energy, Energy Information Administration, Office of Integrated Analysis and Forecasting, January 2004, http://tonto.eia.doe .gov/FTPROOT/forecasting/0383(2004).pdf (accessed November 16, 2004)

FIGURE 3.14

Natural gas consumption by end-use sector, 1990–2025

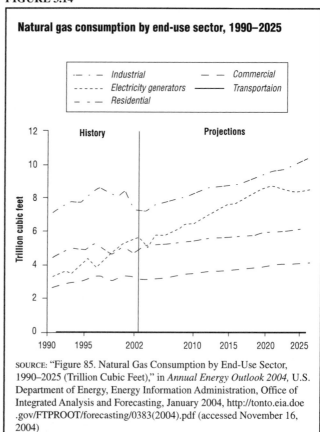

SOURCE: "Figure 85. Natural Gas Consumption by End-Use Sector, 1990–2025 (Trillion Cubic Feet)," in *Annual Energy Outlook 2004,* U.S. Department of Energy, Energy Information Administration, Office of Integrated Analysis and Forecasting, January 2004, http://tonto.eia.doe .gov/FTPROOT/forecasting/0383(2004).pdf (accessed November 16, 2004)

FIGURE 3.15

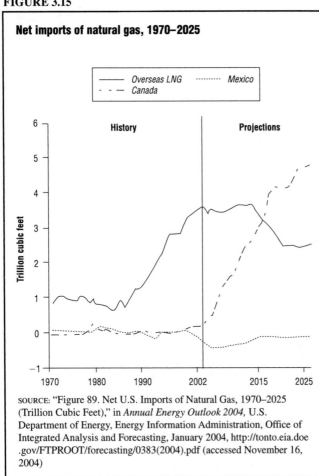

Net imports of natural gas, 1970–2025

SOURCE: "Figure 89. Net U.S. Imports of Natural Gas, 1970–2025 (Trillion Cubic Feet)," in *Annual Energy Outlook 2004,* U.S. Department of Energy, Energy Information Administration, Office of Integrated Analysis and Forecasting, January 2004, http://tonto.eia.doe .gov/FTPROOT/forecasting/0383(2004).pdf (accessed November 16, 2004)

CHAPTER 4
COAL

A HISTORICAL PERSPECTIVE

Although it had been used to create energy for centuries, the first large-scale use of coal occurred during the Industrial Revolution in England from the mid-eighteenth to the mid-nineteenth centuries. At that time the sky was filled with billowing columns of black smoke, soot covered the towns and cities, and workers breathed the thick coal dust swirling around them. Most people then were not concerned with environmental issues because the Industrial Revolution meant jobs to the workers, and factory owners had little desire to control the pollution their factories were creating. In addition, environmental and public health considerations were not as well understood as they are today.

In the United States early colonists used wood to heat their homes because it was so plentiful and coal was less available. Prior to the Civil War (1861–65), some industries used coal as a source of energy, but its major use began with the building of railroads across the country. After the Civil War ended, the United States began to expand its railway system westward and increase its manufacturing capacity. Coal became such a fundamental part of American industrialization that some have called this era the Coal Age. As in England, Americans considered the development of industry a source of national pride. Photographs and postcards of the time proudly featured railroad trains and steel mills with smokestacks belching dark smoke into gray skies.

By the early twentieth century coal had become the major fuel source in the United States, accounting for nearly 90% of the nation's energy requirements. As oil began to heat homes and offices, however, and the growing number of cars used gasoline, coal's dominance declined. By the end of World War II (1939–45), it accounted for only 38% of the energy consumption. Coal fell further out of favor as an energy source in the 1950s and 1960s as oil became more attractive as a cleaner fuel for heating homes and businesses. The decline of coal use

continued, with coal producing as little as 18% of the energy used during some years in the early 1970s because of concerns about environmental pollution and the emergence of nuclear power as a promising energy source.

By 1973, however, Americans recognized they could no longer rely on imported oil for their energy. The OPEC oil embargo clearly demonstrated the nation's heavy reliance on foreign sources of energy and its potentially crippling effect on the American economy. Consequently, the nation revived its interest in domestic coal as a plentiful and economical energy source.

After the 1973 embargo, coal and nuclear fuel received more attention, especially in the electric utility sector. In 1977 President Jimmy Carter called for a two-thirds annual increase in national coal production by 1985. He also asked utility companies and other large industries to convert their operations to coal and proposed a ten-year, $10 billion program to encourage domestic coal production. In 2003 more coal was produced within the United States than any other form of energy, generating 22.3 quadrillion Btu and 32% of all energy produced. (See Figure 1.4 in Chapter 1.) Coal was the second largest source of energy consumed in the United States in 2003, after petroleum.

WHAT IS COAL?

Coal is a black, combustible, mineral solid that develops over millions of years from the partial decomposition of plant matter in an airless space, under increased temperature and pressure. Coal beds, sometimes called seams, are found in the earth between beds of sandstone, shale, and limestone and range in thickness from less than an inch to more than one hundred feet. Approximately five to ten feet of ancient, composted plant material have been compressed to create each foot of coal.

Coal is used as a fuel and in the production of coke (the solid substance left after coal gas and coal tar have been

FIGURE 4.1

Coal-bearing areas

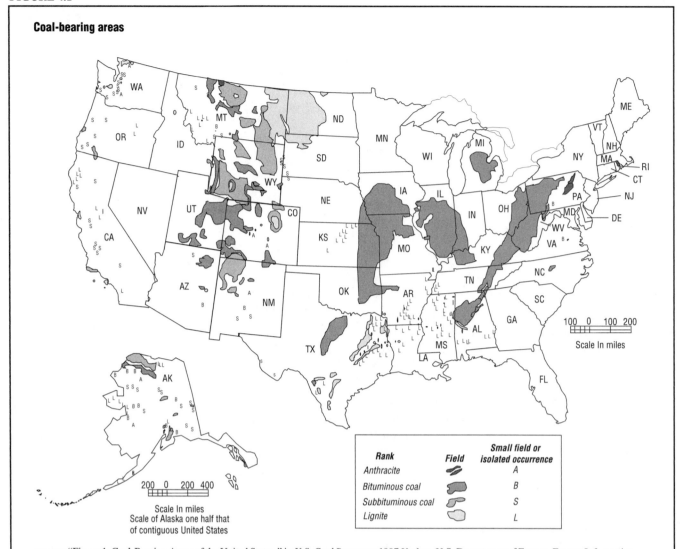

Rank	Field	Small field or isolated occurrence
Anthracite		A
Bituminous coal		B
Subbituminous coal		S
Lignite		L

Scale In miles

Scale of Alaska one half that of contiguous United States

SOURCE: "Figure 1. Coal-Bearing Areas of the United States," in *U.S. Coal Reserves: 1997 Update,* U.S. Department of Energy, Energy Information Administration, 1999, http://www.eia.doe.gov/cneaf/coal/reserves/chapter1.html#fig1 (accessed November 18, 2004)

extracted), coal gas, water gas, and many coal-tar compounds. When coal is burned, its fossil energy—sunlight converted and stored by plants over millions of years—is released. One ton of coal produces an average of 22 million Btu, about the same heating value as 22,000 cubic feet of natural gas, 159 gallons of distillate fuel oil, or one cord of seasoned firewood. (A cord is a stack of wood four feet by four feet by eight feet, or 128 cubic feet.)

CLASSIFICATIONS OF COAL

There are four basic types of coal. Classifications, or "coal ranks," are based on how much carbon, volatile matter, and heating value are contained in the coal.

- Anthracite, or hard coal, is the highest ranked coal. It is hard and jet black, with a moisture content of less than 15%. Anthracite is used mainly for generating electricity and for space heating. It contains approximately 22 to 28 million Btu per ton, with an ignition

temperature of approximately 925 to 970 degrees Fahrenheit. Anthracite is mined mainly in northeastern Pennsylvania. (See Figure 4.1.)

- Bituminous, or soft coal, is the most common coal. It is dense and black, with a moisture content of less than 20% and an ignition range of 700 to 900 degrees Fahrenheit. Bituminous coal is used to generate electricity, for space heating, and to produce coke. Bituminous coal contains a heating value range of 19 to 30 million Btu per ton. It is mined chiefly in the Appalachian and Midwest regions of the United States. (See Figure 4.1.)

- Subbituminous coal, or black lignite, is dull black in color and generally contains 20 to 30% moisture. Black lignite is used for generating electricity and for space heating. It contains 16 to 24 million Btu per ton. Black lignite is mined primarily in the western United States. (See Figure 4.1.)

TABLE 4.1

Coal production, selected years, 1949–2003

(Million short tons)

Year	Rank				Mining method		Location		Total[1]
	Bituminous coal[1]	Subbituminous coal	Lignite	Anthracite[1]	Underground	Surface	East of the Mississippi[1]	West of the Mississippi[1]	
1949	437.9	[2]	[2]	42.7	358.9	121.7	444.2	36.4	480.6
1950	516.3	[2]	[2]	44.1	421.0	139.4	524.4	36.0	560.4
1955	464.6	[2]	[2]	26.2	358.0	132.9	464.2	26.6	490.8
1960	415.5	[2]	[2]	18.8	292.6	141.7	413.0	21.3	434.3
1965	512.1	[2]	[2]	14.9	338.0	189.0	499.5	27.4	527.0
1970	578.5	16.4	8.0	9.7	340.5	272.1	567.8	44.9	612.7
1971	521.3	22.2	8.7	8.7	277.2	283.7	509.9	51.0	560.9
1972	556.8	27.5	11.0	7.1	305.0	297.4	538.2	64.3	602.5
1973	543.5	33.9	14.3	6.8	300.1	298.5	522.1	76.4	598.6
1974	545.7	42.2	15.5	6.6	278.0	332.1	518.1	91.9	610.0
1975	577.5	51.1	19.8	6.2	293.5	361.2	543.7	110.9	654.6
1976	588.4	64.8	25.5	6.2	295.5	389.4	548.8	136.1	684.9
1977	581.0	82.1	28.2	5.9	266.6	430.6	533.3	163.9	697.2
1978	534.0	96.8	34.4	5.0	242.8	427.4	487.2	183.0	670.2
1979	612.3	121.5	42.5	4.8	320.9	460.2	559.7	221.4	781.1
1980	628.8	147.7	47.2	6.1	337.5	492.2	578.7	251.0	829.7
1981	608.0	159.7	50.7	5.4	316.5	507.3	553.9	269.9	823.8
1982	620.2	160.9	52.4	4.6	339.2	499.0	564.3	273.9	838.1
1983	568.6	151.0	58.3	4.1	300.4	481.7	507.4	274.7	782.1
1984	649.5	179.2	63.1	4.2	352.1	543.9	587.6	308.3	895.9
1985	613.9	192.7	72.4	4.7	350.8	532.8	558.7	324.9	883.6
1986	620.1	189.6	76.4	4.3	360.4	529.9	564.4	325.9	890.3
1987	636.6	200.2	78.4	3.6	372.9	545.9	581.9	336.8	918.8
1988	638.1	223.5	85.1	3.6	382.2	568.1	579.6	370.7	950.3
1989	659.8	231.2	86.4	3.3	393.8	586.9	599.0	381.7	980.7
1990	693.2	244.3	88.1	3.5	424.5	604.5	630.2	398.9	1,029.1
1991	650.7	255.3	86.5	3.4	407.2	588.8	591.3	404.7	996.0
1992	651.8	252.2	90.1	3.5	407.2	590.3	588.6	409.0	997.5
1993	576.7	274.9	89.5	4.3	351.1	594.4	516.2	429.2	945.4
1994	640.3	300.5	88.1	4.6	399.1	634.4	566.3	467.2	1,033.5
1995	613.8	328.0	86.5	4.7	396.2	636.7	544.2	488.7	1,033.0
1996	630.7	340.3	88.1	4.8	409.8	654.0	563.7	500.2	1,063.9
1997	653.8	345.1	86.3	4.7	420.7	669.3	579.4	510.6	1,089.9
1998	640.6	385.9	85.8	5.3	417.7	699.8	570.6	547.0	1,117.5
1999	601.7	406.7	87.2	4.8	391.8	708.6	529.6	570.8	1,100.4
2000	574.3	409.2	85.6	4.6	373.7	700.0	507.5	566.1	1,073.6
2001	611.3	434.4	80.0	[1]11.9	380.6	[1]747.1	[1]528.8	[1]598.9	[1]1,127.7
2002	[R]572.1	[R]438.4	[R]82.5	[R]1.4	[R]357.4	[R]736.9	[R]492.9	601.4	[R]1,094.3
2003	[E]559.2	[E]428.4	[E]80.6	[E]1.3	[E]351.2	[E]718.3	[E]468.2	[E]601.3	[P]1,069.5

[1]Beginning in 2001, includes a small amount of refuse recovery.
[2]Included in "Bituminous coal."
R=Revised.
P=Preliminary.
E=Estimate.
Note: Totals may not equal sum of components due to independent rounding.
Web Pages: For data not shown for 1951–1969, see http://www.eia.doe.gov/emeu/aer/coal.html. For related information, see http://www.eia.doe.gov/fuelcoal.html.

SOURCE: "Table 7.2. Coal Production, Selected Years, 1949–2003 (Million Short Tons)," in *Annual Energy Review 2003*, U.S. Department of Energy, Energy Information Administration, Office of Energy Markets and End Use, September 7, 2004, http://www.eia.doe.gov/emeu/aer/pdf/aer.pdf (accessed September 28, 2004)

• Lignite, the lowest ranked coal, is brownish-black in color and has a high moisture content. It tends to disintegrate when exposed to weather. Lignite is used mainly to generate electricity and contains about 9 to 17 million Btu per ton. Lignite has an ignition temperature of approximately 600 degrees Fahrenheit. Most lignite is mined in North Dakota, Montana, Texas, California, and Louisiana. (See Figure 4.1.)

Bituminous coal accounts for the largest share of all coal production; subbituminous is second. (See Table 4.1.) In 2003 production of all types of coal totaled nearly 1.1

billion short tons. (A short ton of coal is 2,000 pounds.) Of that, nearly 1 billion short tons (92%) were bituminous and subbituminous coal. Lignite and anthracite accounted for the remainder.

LOCATIONS OF COAL DEPOSITS

Coal is found in about 13%, or 458,600 square miles, of the total land area of the United States. (See Figure 4.1.) Geologists have divided U.S. coalfields into three geographical zones: the Appalachian, Interior, and Western regions. The Appalachian region is subdivided into three areas: Northern (Ohio, Pennsylvania, Maryland, and northern West Virginia); Central (Virginia, southern West Virginia, eastern Kentucky, and Tennessee); and Southern Appalachia (Alabama). Coal production in the Interior region occurs in Illinois, Indiana, western Kentucky, Iowa, Missouri, Kansas, Arkansas, Oklahoma, Louisiana, and Texas. The Western region includes the Northern Great Plains (Montana, Wyoming, northern Colorado, and North and South Dakota), the Rocky Mountains, the Southwest (southern Colorado, Utah, Arizona, and New Mexico), and the Northwest (Washington and Alaska).

Historically, more coal has been mined east of the Mississippi River than west of the Mississippi, but the West's proportion of total production has increased almost every year since 1965, overtaking the east in 1999. (See Table 4.1.) In 1965 the production of coal in the West was 27 million short tons, only 5% of the national total. By 1999 western production had increased more than twenty-fold, to 570.8 million short tons, or 52% of the total. The amount of coal mined east of the Mississippi that year was 529.6 million short tons. In 2003 slightly more than 601 million short tons of coal were mined west of the Mississippi, while slightly more than 468 million short tons were mined to the east. Western production neared 56% of the total mined.

The growth in coal production in the West has been partly the result of environmental concerns that have led to an increased demand for low-sulfur coal, which is concentrated in the West. In addition, surface mining, which is cheaper and more efficient, is more prevalent in the West. Finally, improved rail service has made it easier to deliver this low-sulfur coal to utility plants located east of the Mississippi River.

COAL MINING METHODS

The method used to mine coal depends on the terrain and the depth of the coal. Prior to the early 1970s, most coal was taken from underground mines. Since that time, however, coal production has shifted from underground mines to surface mines. (See Table 4.1 and Figure 4.2.)

Underground mining is required when the coal lies deeper than 200 feet below ground level. The depth of most underground mines is less than 1,000 feet, but a few go down as far as 2,000 feet. In underground mines some coal must be left untouched in order to form pillars that prevent the mine from caving in. In both underground mines and surface mines, natural features such as folded, faulted, and interlaid rock strata reduce the amount of coal that can be recovered.

Surface mines are usually less than 200 feet deep and can be developed in flat or hilly terrain. Area surface mining is practiced on large plots of relatively flat ground, while contour surface mining follows coal beds along hillsides. (See Figure 4.3.) Open pit mining is used to mine thick, steeply inclined coal beds and uses a combination of contour and area mining methods.

FIGURE 4.2

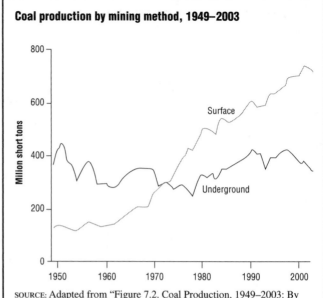

SOURCE: Adapted from "Figure 7.2. Coal Production, 1949–2003: By Mining Method," in *Annual Energy Review 2003,* U.S. Department of Energy, Energy Information Administration, Office of Energy Markets and End Use, September 7, 2004, http://www.eia.doe.gov/emeu/aer/pdf/aer.pdf (accessed September 28, 2004)

The growing prevalence of surface coal mining and the closing of nonproductive mines led to increases in coal mining productivity through the 1980s and 1990s. (See Figure 4.4.) In 2000 average productivity reached an all-time high of seven short tons per miner hour. Productivity dipped a bit in 2001 and 2002, but by 2003 it had nearly returned to the 2000 level. Because surface mines are easier to work, they average up to three times the productivity of underground mines. In 2003 the productivity for surface mines was 10.7 short tons of coal per miner hour, while underground mines produced 4.1 short tons per miner hour, as reported in the *Annual Energy Review 2003,* published in 2004 by the Energy Information Administration (EIA) of the U.S. Department of Energy (DOE).

COAL IN THE DOMESTIC MARKET

Overall Production and Consumption

The EIA noted in its *Annual Energy Review 2003* that the nation consumed 558.4 million short tons of coal in 1974. Twenty-nine years later, in 2003, consumption had grown to nearly 1.1 billion short tons. (Figure 4.5 shows the flow of coal in 2003.) The increases in coal consumption were greatest in the electric utility sector, as many existing electric power plants switched to coal from more expensive oil and gas, and many new, coal-fired power plants were constructed in the 1970s.

Coal Consumption by Sector

To make electricity, coal is pulverized and burned to produce steam, which then drives electric generators.

FIGURE 4.3

Coal mining methods

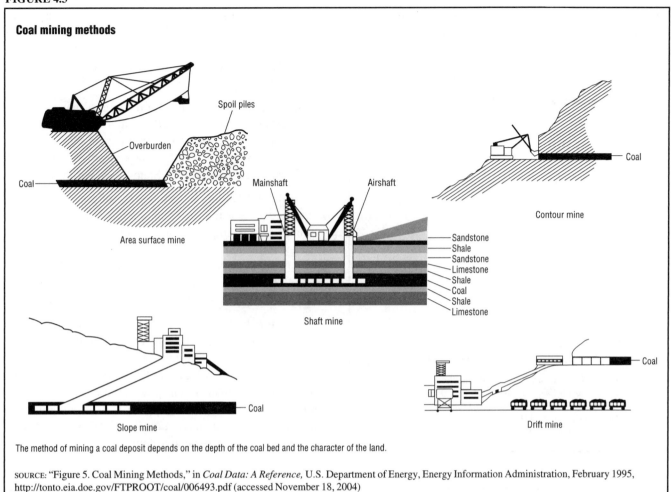

The method of mining a coal deposit depends on the depth of the coal bed and the character of the land.

SOURCE: "Figure 5. Coal Mining Methods," in *Coal Data: A Reference,* U.S. Department of Energy, Energy Information Administration, February 1995, http://tonto.eia.doe.gov/FTPROOT/coal/006493.pdf (accessed November 18, 2004)

Each ton of coal used by an electric generator produces about 2,000 kilowatt-hours of electricity. In household terms each pound of coal produces enough electricity to light ten 100-watt light bulbs for one hour.

Electric utility companies are by far the largest consumers of coal today. (See Figure 4.5 and Figure 4.6.) They accounted for 92% of domestic coal consumption, or 1 billion short tons, in 2003. Coal-fired plants produced nearly 22.3 Btu of electricity, or 40% of U.S. electricity net generation, in 2003. (See Figure 1.5 in Chapter 1.)

The industrial sector was the second-largest consumer of coal in 2003, accounting for 8% of coal use (see Figure 4.6), or 85.4 million short tons. Coal is used in many industrial applications, including the chemical, cement, paper, synthetic fuels, metals, and food-processing industries.

Coal was once a significant fuel source in the residential and commercial sector. (See Figure 4.6.) In 1949 these sectors used 116.5 million short tons of coal. After the late 1940s, however, coal was replaced by oil, natural gas, and electricity, which are cleaner and more convenient. By 1970 only 16.1 million short tons of coal were used in the residential and commercial sectors. Since then, residential

and commercial coal use has continued to decline, falling to 4.5 million short tons in 2003, or far less than 1% of total coal use.

The Price of Coal

In 2003 the average price of coal fell to $17 per short ton, up slightly from the all-time low in 2000 and only 33% of the 1975 price in real dollars, which are adjusted for inflation. (See Table 4.2.) On a per-Btu basis, coal remains the least expensive fossil fuel. In 2000 the average cost of coal was $1.27 per million Btu, compared with $5.68 per million Btu for natural gas and $4.74 per million Btu for residential fuel oil, according to the *Annual Energy Review 2003*.

ENVIRONMENTAL AND HEALTH CONCERNS ABOUT COAL

Problems

The negative side of energy use—pollution of the environment—is not a recent problem. In 1306 King Edward I of England so objected to the noxious smoke from London's coal-burning fires that he banned coal's use by everyone except blacksmiths. The enormous scale of today's energy use has increased environmental concerns.

Coal-fired electric power plants emit gases that are harmful to the environment. Scientists believe that burning huge quantities of fossil fuels causes the "greenhouse effect," in which gases from the fuels trap heat in the earth's atmosphere and cause increased warming, which threatens the environment. Burning coal also contributes to the formation of acid rain and to public health concerns. Sulfur dioxide, for instance, has been shown to cause respiratory problems.

Carbon dioxide accounts for the largest share of greenhouse gas emissions. In 2002 the combustion of coal in the United States produced 2.1 billion metric tons of carbon dioxide, or 36% of total carbon dioxide emissions from all fossil fuels used in the United States. (See Figure 4.7.)

ACID RAIN. Acid rain is any form of precipitation that contains a greater-than-normal amount of acid. Even non-polluted rain is slightly acidic (with a pH of about 5.6) because rainwater combines with the carbon dioxide normally found in the air to produce a weak acid called carbonic acid. But pollutants in the air can increase the acidity of rain and other forms of precipitation, such as snow and fog.

Chemicals such as oxides of sulfur and nitrogen, which are given off during the combustion of fossil fuels, are pollutants that combine with precipitation to form acids. These oxides increase in the air because of automobile exhaust, industrial and power plant emissions, and other fossil fuel combustion processes. In many parts of the world acid rain has caused significant damage to forests, lakes, and other ecosystems.

FIGURE 4.4

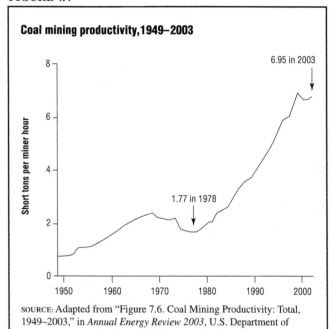

Coal mining productivity, 1949–2003

SOURCE: Adapted from "Figure 7.6. Coal Mining Productivity: Total, 1949–2003," in *Annual Energy Review 2003*, U.S. Department of Energy, Energy Information Administration, Office of Energy Markets and End Use, September 7, 2004, http://www.eia.doe.gov/emeu/aer/pdf/aer.pdf (accessed September 28, 2004)

FIGURE 4.5

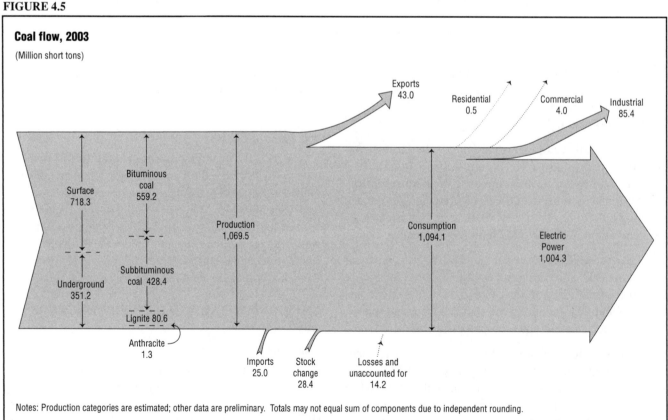

Coal flow, 2003

(Million short tons)

Notes: Production categories are estimated; other data are preliminary. Totals may not equal sum of components due to independent rounding.

SOURCE: "Diagram 4. Coal Flow, 2003 (Million Short Tons)," in *Annual Energy Review 2003,* U.S. Department of Energy, Energy Information Administration, Office of Energy Markets and End Use, September 7, 2004, http://www.eia.doe.gov/emeu/aer/pdf/aer.pdf (accessed September 28, 2004)

TABLE 4.2

Coal prices, selected years, 1949–2003

(Dollars per short ton)

Year	Bituminous Coal		Subbituminous Coal		Lignite[1]		Anthracite		Total	
	Nominal	Real[2]	Nominal	Real[2]	Nominal	Real[2]	Nominal	Real[2]	Nominal	Real[2]
1949	[3]4.90	[3,R]29.97	[3]	[3]	2.37	[R]14.49	8.90	[R]54.43	5.24	[R]32.05
1950	[3]4.86	[3,R]29.40	[3]	[3]	2.41	[R]14.58	9.34	[R]56.50	5.19	[R]31.40
1955	[3]4.51	[3,R]24.06	[3]	[3]	2.38	[R]12.70	8.00	[R]42.68	4.69	[R]25.02
1960	[3]4.71	[3,R]22.38	[3]	[3]	2.29	[R]10.88	8.01	[R]38.07	4.83	[R]22.96
1965	[3]4.45	[3,R]19.75	[3]	[3]	2.13	[R]9.45	8.51	[R]37.76	4.55	[R]20.19
1970	[3]6.30	[3,R]22.88	[3]	[3]	1.86	[R]6.76	11.03	[R]40.06	6.34	[R]23.03
1971	[3]7.13	[3,R]24.66	[3]	[3]	1.93	[R]6.68	12.08	[R]41.78	7.15	[R]24.73
1972	[3]7.78	[3,R]25.79	[3]	[3]	2.04	[R]6.76	12.40	[R]41.11	7.72	[R]25.59
1973	[3]8.71	[3,R]27.35	[3]	[3]	2.09	[R]6.56	13.65	[R]42.86	8.59	[R]26.97
1974	[3]16.01	[3,R]46.11	[3]	[3]	2.19	[R]6.31	22.19	[R]63.90	15.82	[R]45.56
1975	[3]19.79	[3,R]53.08	[3]	[3]	3.17	[R]8.34	32.26	[R]84.89	19.35	[R]50.92
1976	[3]20.11	[3,R]50.03	[3]	[3]	3.74	[R]9.30	33.92	[R]84.39	19.56	[R]48.66
1977	[3]20.59	[3,R]48.16	[3]	[3]	4.03	[R]9.43	34.86	[R]81.54	19.95	[R]46.66
1978	[R]22.64	[3,R]48.48	[3]	[3]	5.68	[R]12.41	35.25	[R]77.04	21.86	[R]47.77
1979	27.31	[R]55.12	9.55	[R]19.27	6.48	[R]13.08	41.06	[R]82.87	23.75	[R]47.93
1980	29.17	[R]53.98	11.08	[R]20.50	7.60	[R]14.06	42.51	[R]78.66	24.65	[R]45.61
1981	31.51	[R]53.30	12.18	[R]20.60	8.85	[R]14.97	44.28	[R]74.90	26.40	[R]44.66
1982	32.15	[R]51.25	13.37	[R]21.31	9.79	[R]15.61	49.85	[R]79.47	27.25	[R]43.44
1983	31.11	[R]47.71	13.03	[R]19.98	9.91	[R]15.20	52.29	[R]80.19	25.98	[R]39.84
1984	30.63	[R]45.27	12.41	[R]18.34	10.45	[R]15.45	48.22	[R]71.27	25.61	[R]37.85
1985	30.78	[R]44.15	12.57	[R]18.03	10.68	[R]15.32	45.80	[R]65.70	25.20	[R]36.15
1986	28.84	[R]40.48	12.26	[R]17.21	10.64	[R]14.93	44.12	[R]61.92	23.79	[R]33.39
1987	28.19	[R]38.51	11.32	[R]15.47	10.85	[R]14.82	43.65	[R]59.63	23.07	[R]31.52
1988	27.66	[R]36.54	10.45	[R]13.81	10.06	[R]13.29	44.16	[R]58.34	22.07	[R]29.16
1989	27.40	[R]34.88	10.16	[R]12.93	9.91	[R]12.62	42.93	[R]54.65	21.82	[R]27.78
1990	27.43	[R]33.62	9.70	[R]11.89	10.13	[R]12.42	39.40	[R]48.29	21.76	[R]26.67
1991	27.49	[R]32.55	9.68	[R]11.46	10.89	[R]12.90	36.34	[R]43.03	21.49	[R]25.45
1992	26.78	[R]31.00	9.68	[R]11.21	10.81	[R]12.51	34.24	[R]39.64	21.03	[R]24.34
1993	26.15	[R]29.59	9.33	[R]10.56	11.11	[R]12.57	32.94	[R]37.27	19.85	[R]22.46
1994	25.68	[R]28.45	8.37	[R]9.27	10.77	[R]11.93	36.07	[R]39.96	19.41	[R]21.50
1995	25.56	[R]27.75	8.10	[R]8.79	10.83	[R]11.76	39.78	[R]43.19	18.83	[R]20.44
1996	25.17	[R]26.82	7.87	[R]8.39	10.92	[R]11.64	36.78	[R]39.19	18.50	[R]19.71
1997	24.64	[R]25.82	7.42	[R]7.78	10.91	[R]11.43	35.12	[R]36.81	18.14	[R]19.01
1998	24.87	[R]25.78	6.96	[R]7.21	11.08	[R]11.49	42.91	[R]44.48	17.67	[R]18.32
1999	23.92	[R]24.44	6.87	[R]7.02	11.04	[R]11.28	35.13	[R]35.90	16.63	[R]16.99
2000	24.15	[R]24.15	7.12	[R]7.12	11.41	[R]11.41	40.90	[R]40.90	16.78	[R]16.78
2001	25.36	[R]24.77	6.67	[R]6.52	11.52	[R]11.25	47.67	[R]46.57	17.38	[R]16.98
2002	[R]26.57	[R]25.56	[R]7.34	[R]7.06	[R]11.07	[R]10.65	[R]47.78	[R]45.97	[R]17.98	[R]17.30
2003[E]	26.57	25.14	7.34	6.95	11.07	10.48	47.78	45.21	17.98	17.01

[1]Because of withholding to protect company confidentiality, lignite prices exclude Texas for 1955–1977 and Montana for 1974–1978. As a result, lignite prices for 1974–1977 are North Dakota only.
[2]In chained (2000) dollars, calculated by using gross domestic product implicit price deflators.
[3]Through 1978, subbituminous cola included in "Bituminous coal."
R = Revised.
E = Estimate.
Note: Prices are free-on-board (f.o.b.) rail/barge prices, which are the f.o.b. prices of coal at the point of first sale, excluding freight or shipping and insurance costs.
Web Pages: For data not shown for 1951–1969, see http://www.eia.doe.gov/emeu/aer/coal.html. For related information, see http://www.eia.doe.gov/fuelcoal.html.

SOURCE: "Table 7.8. Coal Prices, Selected Years, 1949–2003 (Dollars per Short Ton)," in *Annual Energy Review 2003*, U.S. Department of Energy, Energy Information Administration, Office of Energy Markets and End Use, September 7, 2004, http://www.eia.doe.gov/emeu/aer/pdf/aer.pdf (accessed September 28, 2004)

HEALTH ISSUES. Emissions from coal-fired power plants include mercury, sulfur oxides, and nitrogen oxides. Mercury reaches humans when they eat fish contaminated by airborne mercury that settles in lakes and streams. Scientific data have not yet determined a significant link between mercury that originates from coal-fired power plants and significant health effects in humans. However, sulfur oxides and nitrogen oxides contribute to air pollution, which can cause upper respiratory conditions (see Table 4.3.)

Coal miners are at risk for developing pneumoconiosis (black lung disease), which results from chronic inhalation of coal dust. This risk has been drastically reduced, howev-er, by using personal protective equipment such as dust masks and respirators, covering the walls of tunnels and shafts with pulverized white rock to lower the level of the dust, and spraying water to promote settling of the dust.

Solutions

THE CLEAN COAL TECHNOLOGY LAW. In 1984 Congress established the DOE's Clean Coal Technology (CCT) program (PL 98–473). Congress directed the DOE to administer cost-shared projects (financed by both industry and government) to demonstrate clean coal technologies. The demonstration projects had the goal of

FIGURE 4.6

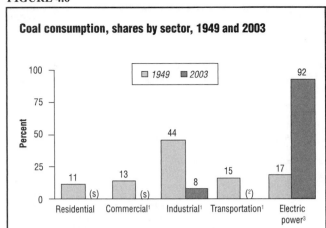

Coal consumption, shares by sector, 1949 and 2003

¹Includes combined-heat-and-power plants and a small number of electricity-only plants.
²For 1978 forward, small amouts of transportation sector use are included in "Industrial."
³Electricity-only and combined-heat-and-power plants whose primary business is to sell electricity, or electricity and heat, to the public.
(s)=Less than 0.5 million short tons or less than 0.5 percent, as appropriate.

SOURCE: Adapted from "Figure 7.3. Coal Consumption by Sector: Sector Shares, 1949 and 2003," in *Annual Energy Review 2003*, U.S. Department of Energy, Energy Information Administration, Office of Energy Markets and End Use, September 7, 2004, http://www.eia.doe.gov/emeu/aer/pdf/aer.pdf (accessed September 28, 2004)

TABLE 4.3

Air pollutants, health risks, and contributing sources

Pollutants	Health risks	Contributing sources
Ozone¹ (O_3)	Asthma, reduced respiratory function, eye irritation	Cars, refineries, dry cleaners
Particulate matter (PM-IO)	Bronchitis, cancer, lung damage	Dust, pesticides
Carbon monoxide (CO)	Blood oxygen carrying capacity reduction, cardiovascular and nervous system impairments	Cars, power plants, wood stoves
Sulphur dioxide (SO_2)	Respiratory tract impairment, destruction of lung tissue	Power plants, paper mills
Lead (Pb)	Retardation and brain damage, esp. children	Cars, nonferrous smelters, battery plants
Nitrogen dioxide (NO_2)	Lung damage and respiratory illness	Power plants, cars, trucks

¹Ozone refers to tropospheric ozone which is hazardous to human health.

SOURCE: Fred Seitz and Christine Plepys, "Table 1. Criteria Air Pollutants, Health Risks and Sources," in *Healthy People 2000: Statistical Notes, Number 9*, Centers for Disease Control and Prevention, National Center for Health Statistics, September 1995, http://www.cdc.gov/nchs/data/statnt/statnt09.pdf (accessed November 21, 2004)

---

using coal in more environmentally and economically efficient ways.

CLEAN COAL TECHNOLOGY AND THE CLEAN AIR ACT. The stated goal of both Congress and the DOE has been to

---

FIGURE 4.7

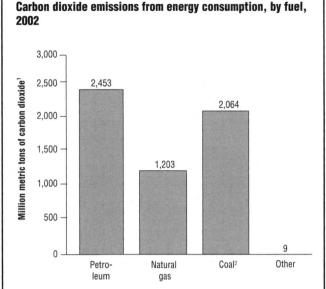

Carbon dioxide emissions from energy consumption, by fuel, 2002

¹Metric tons of carbon dioxide can be converted to metric tons of carbon equivalent by multiplying by 12/44.
²Coal coke net imports, municipal solid waste, and geothermal.

SOURCE: Adapted from "Figure 12.3. Carbon Dioxide Emissions from Energy Consumption by Sector by Energy Source, 2002: Total by Fuel," in *Annual Energy Review 2003*, U.S. Department of Energy, Energy Information Administration, Office of Energy Markets and End Use, September 7, 2004, http://www.eia.doe.gov/emeu/aer/pdf/aer.pdf (accessed September 28, 2004)

---

develop cost-effective ways to burn coal more cleanly, both to control acid rain and to improve the nation's energy security by reducing dependence on imported fuels. One strategy is a slow, phased-in approach in which utility companies and states reduce their emissions in stages.

Under the Clean Air Act of 1990 (PL 101–549), restrictions on sulfur dioxide and nitrogen oxide emissions took effect in 1995 and tightened in 2000. Each round of regulation requires coal-burning utilities to find lower-sulfur coal or to install cleaner technology, such as "scrubbers" that reduce smokestack emissions that contribute heavily to air pollution. When the first Clean Air Act was passed in 1970, it was aimed at changing the air-quality standards at new generating stations, and older coal-using plants were exempt. Under the 1990 act, older plants are also covered by the regulations.

In January 2004 the Environmental Protection Agency (EPA) proposed new regulations for reducing emissions of sulfur dioxide, nitrogen oxides, and mercury from coal-burning power plants. The Interstate Air Quality Rule focuses on twenty-nine eastern states whose sulfur dioxide and nitrogen oxide emissions are significantly contributing to fine particle and ozone pollution problems. The Utility Mercury Reductions Rule focuses on controlling mercury emissions from power plants. (When taken into the body, mercury can result in serious health effects in children.) These proposed actions strengthen Clean Air Act regulations and standards but are not specifically mandated by the Congress. When implemented, the EPA expects them to result in rapid and significant air quality improvement.

CLEANER COAL USE. The coal-burning process can be cleaned by physical or chemical methods. Scrubbers, which are a physical method commonly used to reduce sulfur dioxide emissions, filter coal emissions by spraying lime or a calcium compound and water across the emission stream before it leaves the smokestack. The sulfur dioxide bonds to the spray and settles as a mudlike substance that can be pumped out for disposal. Scrubbers, however, are expensive to operate, so particulate collectors are the most common emissions cleaners for coal. While they are cheaper to operate than scrubbers, they are less effective. Cooling towers reduce heat released into the atmosphere and reduce some pollutants. Chemical cleaning, a relatively new technology not yet in widespread use, involves the use of biological or chemical agents to clean emissions.

Under the new environmental regulations of the 1990 Clean Air Act and its amendments, plants with coal-generated boilers must be built to reduce sulfur emissions by 70 to 90%. New, high-sulfur coal electricity plants, designed to meet emission standards, use 30% of their construction costs on pollution control equipment and take up to 5% of their power output to operate this equipment. Research to lower these costs is important because of the quantity of electricity produced with coal in the United States.

The EIA's *Annual Energy Review 2003* (2004) notes that in 2002 coal-fired electricity plants that had environmental equipment installed had a production capacity of 329.5 gigawatts (1 gigawatt equals 1,000 megawatts). Of this capacity, 100% was generated within plants using particulate collectors, 47% in those with cooling towers, and 30% in those with scrubbers. (Some plants use more than type of environmental equipment so the figures add up to more than 100%.) The use of scrubbers is projected to increase as new regulations from the 1990 Clean Air Act and its amendments take effect.

COAL EXPORTS

Since 1950 the United States has produced more coal than it has consumed. The excess production has allowed the United States to become a significant exporter of coal to other nations. However, exports of this energy source have declined dramatically since 1991, when the U.S. exported 109 million short tons of coal. In 2003 the U.S. exported 43 million short tons, up from 39.6 million short tons in 2002, the lowest amount exported since 1961. (See Table 4.4.)

In 2003 coal made up 28% of all U.S. energy exports. (See Figure 1.5 in Chapter 1.) Europe received 35% of U.S. coal exports. The individual countries that bought the most U.S. coal were Canada, Brazil, Italy, and the Netherlands. (See Table 4.4.)

INTERNATIONAL COAL USAGE

The *Annual Energy Review 2003* (2004) reported that world coal production was 1.1 billion short tons in 2002 and accounted for 21% of world energy production. China led the world in coal production, mining just over 1.5 billion short tons, followed by the United States at 1.1 billion short tons. (See Figure 4.8.) Other major producers were India, Australia, Russia, South Africa, Germany, and Poland.

World consumption of coal in 2002 totaled 5.3 billion short tons. Besides being the largest producer, China was

FIGURE 4.8

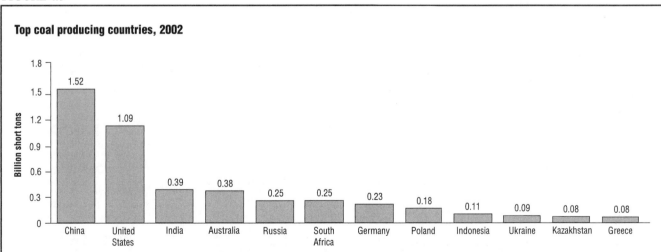

SOURCE: Adapted from "Figure 11.14. World Coal Production: Top Producing Countries, 2002," in *Annual Energy Review 2003,* U.S. Department of Energy, Energy Information Administration, Office of Energy Markets and End Use, September 7, 2004, http://www.eia.doe.gov/emeu/aer/pdf/aer.pdf (accessed September 28, 2004)

TABLE 4.4

Coal exports by country of destination, 1960–2003

(Million short tons)

Year	Canada	Brazil	Europe Belgium and Luxembourg	Denmark	France	Germany[1]	Italy	Netherlands	Spain	United Kingdom	Other	Total	Japan	Other	Total
1960	12.8	1.1	1.1	0.1	0.8	4.6	4.9	2.8	0.3	0.0	2.4	17.1	5.6	1.3	38.0
1961	12.1	1.0	1.0	0.1	0.7	4.3	4.8	2.6	0.2	0.0	2.0	15.7	6.6	1.0	36.4
1962	12.3	1.3	1.3	(s)	0.9	5.1	6.0	3.3	0.8	(s)	1.8	19.1	6.5	1.0	40.2
1963	14.6	1.2	2.7	(s)	2.7	5.6	7.9	5.0	1.5	0.0	2.4	27.7	6.1	0.9	50.4
1964	14.8	1.1	2.3	(s)	2.2	5.2	8.1	4.2	1.4	0.0	2.6	26.0	6.5	1.1	49.5
1965	16.3	1.2	2.2	(s)	2.1	4.7	9.0	3.4	1.4	(s)	2.3	25.1	7.5	0.9	51.0
1966	16.5	1.7	1.8	(s)	1.6	4.9	7.8	3.2	1.2	(s)	2.5	23.1	7.8	1.0	50.1
1967	15.8	1.7	1.4	0.0	2.1	4.7	5.9	2.2	1.0	0.0	2.1	19.4	12.2	1.0	50.1
1968	17.1	1.8	1.1	0.0	1.5	3.8	4.3	1.5	1.5	0.0	1.9	15.5	15.8	0.9	51.2
1969	17.3	1.8	0.9	0.0	2.3	3.5	3.7	1.6	1.8	0.0	1.3	15.2	21.4	1.2	56.9
1970	19.1	2.0	1.9	0.0	3.6	5.0	4.3	2.1	3.2	(s)	1.8	21.8	27.6	1.2	71.7
1971	18.0	1.9	0.8	0.0	3.2	2.9	2.7	1.6	2.6	1.7	1.1	16.6	19.7	1.1	57.3
1972	18.7	1.9	1.1	0.0	1.7	2.4	3.7	2.3	2.1	2.4	1.2	16.9	18.0	1.2	56.7
1973	16.7	1.6	1.2	0.0	2.0	1.6	3.3	1.8	2.2	0.9	1.3	14.4	19.2	1.6	53.6
1974	14.2	1.3	1.1	0.0	2.7	1.5	3.9	2.6	2.0	1.4	0.9	16.1	27.3	1.8	60.7
1975	17.3	2.0	0.6	0.0	3.6	2.0	4.5	2.1	2.7	1.9	1.6	19.0	25.4	2.6	66.3
1976	16.9	2.2	2.2	(s)	3.5	1.0	4.2	3.5	2.5	0.8	2.1	19.9	18.8	2.1	60.0
1977	15.7	2.3	1.5	0.1	2.1	0.9	4.1	2.0	1.6	0.6	2.1	15.0	15.9	3.5	54.3
1978	15.7	1.5	1.1	0.0	1.7	0.6	3.2	1.1	0.8	0.4	2.2	11.0	10.1	2.5	40.7
1979	19.5	2.8	3.2	0.2	3.9	2.6	5.0	2.0	1.4	1.4	4.4	23.9	15.7	4.1	66.0
1980	17.5	3.3	4.6	1.7	7.8	2.5	7.1	4.7	3.4	4.1	6.0	41.9	23.1	6.0	91.7
1981	18.2	2.7	4.3	3.9	9.7	4.3	10.5	6.8	6.4	2.3	8.8	57.0	25.9	8.7	112.5
1982	18.6	3.1	4.8	2.8	9.0	2.3	11.3	5.9	5.6	2.0	7.6	51.3	25.8	7.5	106.3
1983	17.2	3.6	2.5	1.7	4.2	1.5	8.1	4.2	3.3	1.2	6.4	33.1	17.9	6.1	77.8
1984	20.4	4.7	3.9	0.6	3.8	0.9	7.6	5.5	2.3	2.9	5.3	32.8	16.3	7.2	81.5
1985	16.4	5.9	4.4	2.2	4.5	1.1	10.3	6.3	3.5	2.7	10.3	45.1	15.4	9.9	92.7
1986	14.5	5.7	4.4	2.1	5.4	0.8	10.4	5.6	2.6	2.9	8.4	42.6	11.4	11.4	85.5
1987	16.2	5.8	4.6	0.9	2.9	0.5	9.5	4.1	2.5	2.6	6.6	34.2	11.1	12.3	79.6
1988	19.2	5.3	6.5	2.8	4.3	0.7	11.1	5.1	2.5	3.7	8.5	45.1	14.1	11.3	95.0
1989	16.8	5.7	7.1	3.2	6.5	0.7	11.2	6.1	3.3	4.5	8.9	51.6	13.8	12.9	100.8
1990	15.5	5.8	8.5	3.2	6.9	1.1	11.9	8.4	3.8	5.2	9.5	58.4	13.3	12.7	105.8
1991	11.2	7.1	7.5	4.7	9.5	1.7	11.3	9.6	4.7	6.2	10.4	65.5	12.3	13.0	109.0
1992	15.1	6.4	7.2	3.8	8.1	1.0	9.3	9.1	4.5	5.6	8.5	57.3	12.3	11.4	102.5
1993	8.9	5.2	5.2	0.3	4.0	0.5	6.9	5.6	4.1	4.1	6.9	37.6	11.9	11.0	74.5
1994	9.2	5.5	4.9	0.5	2.9	0.3	7.5	4.9	4.1	3.4	7.3	35.8	10.2	10.7	71.4
1995	9.4	6.4	4.5	2.1	3.7	2.0	9.1	7.3	4.7	4.7	10.7	48.6	11.8	12.4	88.5
1996	12.0	6.5	4.6	1.3	3.9	1.1	9.2	7.1	4.1	6.2	9.8	47.2	10.5	14.2	90.5
1997	15.0	7.5	4.3	0.4	3.4	0.9	7.0	4.8	4.1	7.2	9.2	41.3	8.0	11.8	83.5
1998	20.7	6.5	3.2	0.3	3.2	1.2	5.3	4.5	3.2	5.9	6.9	33.8	7.7	9.4	78.0
1999	19.8	4.4	2.1	0.0	2.5	0.6	4.0	3.4	2.5	3.2	4.3	22.5	5.0	6.7	58.5
2000	18.8	4.5	2.9	0.1	3.0	1.0	3.7	2.6	2.7	3.3	5.7	25.0	4.4	5.8	58.5
2001	17.6	4.6	2.8	0.0	2.2	0.9	5.4	2.1	1.6	2.5	3.3	20.8	2.1	3.6	48.7
2002	16.7	3.5	2.4	0.0	1.3	1.0	3.1	1.7	1.9	1.9	2.4	15.6	1.3	2.6	39.6
2003	20.8	3.5	1.8	0.3	1.3	0.5	2.8	2.0	1.8	1.5	3.2	15.1	(s)	3.6	43.0

TABLE 4.4

Coal exports by country of destination, 1960–2003 [CONTINUED]

(Million short tons)

¹Through 1990, data for Germany are for the former West Germany only. Beginning in 1991, data for Germany are for the unified Germany, i.e., the former East Germany and West Germany.
(s) = Less than 0.05 million short tons.
Note: Totals may not equal sum of components due to independent rounding.

SOURCE: "Table 7.4. Coal Exports by Country of Destination, 1960–2003 (Million Short Tons)," in *Annual Energy Review 2003*, U.S. Department of Energy, Energy Information Administration, Office of Energy Markets and End Use, September 7, 2004, http://www.eia.doe.gov/emeu/aer/pdf/aer.pdf (accessed September 28, 2004)

FIGURE 4.9

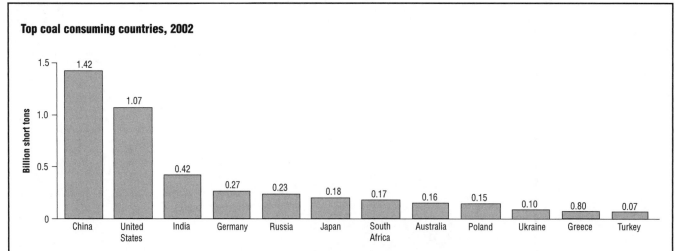

Top coal consuming countries, 2002

SOURCE: Adapted from "Figure 11.15. World Coal Consumption: Top Consuming Countries, 2002," in *Annual Energy Review 2003*, U.S. Department of Energy, Energy Information Administration, Office of Energy Markets and End Use, September 7, 2004, http://www.eia.doe.gov/emeu/aer/pdf/aer.pdf (accessed September 28, 2004)

also the largest consumer of coal in 2002, using more than 1.4 billion short tons, followed by the United States at 1.1 billion short tons. (See Figure 4.9.) Other major consumers included India, Germany, Russia, and Japan.

FUTURE TRENDS IN THE COAL INDUSTRY

In *Annual Energy Outlook 2004* the EIA forecasted that domestic coal production will increase to almost 1.3 billion short tons by 2015 and increase slightly from there to almost 1.4 billion short tons by 2020 and approximately 1.5 billion short tons by 2025. (See Table 4.5.) Domestic consumption is projected to match production, reaching nearly 1.3 billion short tons in 2015, almost 1.4 billion short tons by 2020, and approximately 1.5 billion short tons by 2025. Electricity generation will still use the majority of coal in 2015 (approximately 1.2 billion short tons), 2020, and 2025. Total coal consumption for electricity generation is expected to increase by an average of 1.8% per year because of high natural gas prices.

Coal prices are projected to decline from $17.90 per short ton in 2002 (see Table 4.5) to a low of $16.19 per short ton in 2016. These price decreases will likely occur because of improvements in mine productivity, a shift to western production, declines in rail transportation costs, and competitive pressures on labor costs. After 2016 prices will likely rise as productivity improvements slow and the industry faces increasing costs to open new mining areas. By 2025 the price of coal is projected to rise to $16.57 per short ton. (See Table 4.5.) Nevertheless, this price is still below the 2002 price.

Environmental concerns about acid rain and global warming may continue to grow. The outlook for the U.S. coal industry could be affected by acid rain legislation, the development of clean coal technologies, and, over the longer term, the problem of global warming. Environmental issues will increasingly become international problems. China, with nearly five times the population of the United States and a growing economy, may surpass the United States in carbon emissions by 2020.

The EIA predicts that U.S. coal exports will decline from 6% in 2002 to less than 3% in 2025. This decline will likely be the result of a decline in the demand for coal in Europe and the Americas. Also, other countries are expected to reduce costs and gain currency exchange advantages against the U.S. dollar.

TABLE 4.5

Comparison of coal forecasts, 2015, 2020, and 2025

(Million short tons, except where noted)

Projection	2002	AEO2004			Other forecasts	
		Reference	Low economic growth	High economic growth	EVA	Hill & Associates
2015						
Production	**1,105**	**1,285**	**1,262**	**1,288**	**1,114**	**1,204**
Consumption by sector						
Electricity generation	976	1,200	1,180	1,200	1,042	1,144
Coking plants	23	21	21	21	18	18
Industrial/other	67	70	67	73	60	62
Total	**1,066**	**1,291**	**1,269**	**1,295**	**1,120**	**1,224**
Net coal exports	**22.7**	**−6.1**	**−6.1**	**−6.1**	**−6.2**	**−20.4**
Exports	39.6	31.6	31.6	31.6	29.5	28.4
Imports	16.9	37.7	37.7	37.7	35.7	48.8
Minemouth price						
(2002 dollars per short ton)	17.90	16.47	15.84	16.75	17.02[1]	17.78[2,3]
(2002 dollars per million Btu)	0.87	0.81	0.78	0.82	0.83[1]	0.81[2,3]
Average delivered price to electricity generators						
(2002 dollars per short ton)	25.96	24.34	23.17	25.10	NA	21.82[3]
(2002 dollars per million Btu)	1.26	1.22	1.16	1.25	NA	1.08[3]
2020						
Production	**1,105**	**1,377**	**1,337**	**1,382**	**1,159**	**1,208**
Consumption by sector						
Electricity generation	976	1,301	1,263	1,305	1,095	1,158
Coking plants	23	19	19	19	17	17
Industrial/other	67	71	68	75	57	59
Total	**1,066**	**1,391**	**1,349**	**1,399**	**1,169**	**1,234**
Net coal exports	**22.7**	**−14.4**	**−12.2**	**−15.7**	**−10.4**	**−25.6**
Exports	39.6	27.4	29.5	26.0	29.7	22.6
Imports	16.9	41.7	41.7	41.7	40.1	48.2
Minemouth price						
(2002 dollars per short ton)	17.90	16.32	15.78	16.92	16.91[1]	16.94[2,3]
(2002 dollars per million Btu)	0.87	0.80	0.78	0.83	0.83[1]	0.77[2,3]
Average delivered price to electricity generators						
(2002 dollars per short ton)	25.96	24.01	22.87	25.03	NA	21.08[3]
(2002 dollars per million Btu)	1.26	1.20	1.15	1.24	NA	1.04[3]
2025						
Production	**1,105**	**1,543**	**1,420**	**1,586**	**1,237**	**NA**
Consumption by sector						
Electricity generation	976	1,477	1,355	1,510	1,184	NA
Coking plants	23	17	17	17	16	NA
Industrial/other	67	72	68	84	55	NA
Total	**1,066**	**1,567**	**1,441**	**1,612**	**1,254**	**NA**
Net coal exports	**22.7**	**−22.7**	**−19.8**	**−24.8**	**−17.8**	**NA**
Exports	39.6	23.0	26.0	21.0	30.0	NA
Imports	16.9	45.7	45.7	45.7	47.8	NA
Minemouth price						
(2002 dollars per short ton)	17.90	16.57	15.67	17.95	16.97[1]	NA
(2002 dollars per million Btu)	0.87	0.82	0.78	0.88	0.84[1]	NA
Average delivered price to electricity generators						
(2002 dollars per short ton)	25.96	24.31	22.75	26.29	NA	NA
(2002 dollars per million Btu)	1.26	1.22	1.14	1.30	NA	NA

[1]The average coal price is a weighted average of the projected spot market FOB mine price for all domestic coal.
[2]The minemouth price represents an average for domestics team coal only. Exports and coking coal are not included in the average.
[3]The prices provided by Hill & Associates were converted from 2003 dollars to 2002 dollars in order to be consistent with AEO 2004.
Btu = British thermal unit.
NA = Not available.

SOURCE: "Table 33. Comparison of Coal Forecasts, 2015, 2020, and 2025 (Million Short Tons, except Where Noted)," in *Annual Energy Outlook 2004*, U.S. Department of Energy, Energy Information Administration, Office of Integrated Analysis and Forecasting, January 2004, http://tonto.eia.doe.gov/FTPROOT/forecasting/0383(2004).pdf (accessed November 16, 2004)

CHAPTER 5
NUCLEAR ENERGY

Nuclear energy is used in the United States to generate electricity and to power some navy ships. In the decades since the first commercial nuclear reactor went into operation in 1956, the nuclear power industry has had a difficult time persuading the American public of the safety of its enterprise. In the early 1970s most Americans favored the use of nuclear power because it appeared to provide cleaner, more efficient, energy than fossil fuels, and it could help reduce U.S. dependence on foreign energy sources. But by the early 2000s many people in the United States—and, indeed, around the world—opposed building additional nuclear power plants. The 1979 disaster at Three Mile Island in Pennsylvania and the 1986 catastrophe at Chernobyl in the former Soviet Union greatly increased public concerns about the safety of nuclear power, as have reports of design flaws, cracks, and leaks in other reactors. Furthermore, the safe disposal of radioactive waste, which is a by-product of nuclear energy, has proven difficult.

Supporters of nuclear power believe that it is as safe as any other form of energy production if monitored correctly. They point to growing concerns about fossil fuel use, including global warming and acid rain, as well as damage caused by mining and transporting fossil fuels. In fact, this growing concern over fossil fuels has led a small number of environmentalists who had previously opposed nuclear power to reconsider their position. Nonetheless, environmental, safety, and economic concerns have restrained growth in the nuclear industry since the mid-1970s. Unwillingness to commission new nuclear plants is evident in Figure 5.1, which shows a general leveling off of nuclear energy's share of electricity production from 1988 through 2003 at about 20%.

HOW NUCLEAR ENERGY WORKS

In a nuclear power plant, fuel (uranium in the United States) in the reactor generates a nuclear reaction (fission) that produces heat. In a pressurized water reactor (see Fig-

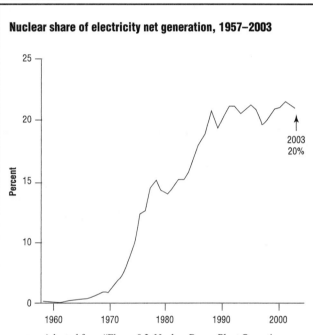

FIGURE 5.1

Nuclear share of electricity net generation, 1957–2003

SOURCE: Adapted from "Figure 9.2. Nuclear Power Plant Operations: Nuclear Share of Electricity Net Generation, 1957–2003," in *Annual Energy Review 2003,* U.S. Department of Energy, Energy Information Administration, Office of Energy Markets and End Use, September 7, 2004, http://www.eia.doe.gov/emeu/aer/pdf/aer.pdf (accessed September 28, 2004)

ure 5.2), the heat from the reaction is carried away by water under high pressure, which heats a second water stream, producing steam. The steam runs through a turbine (similar to a jet engine), making it and the attached electrical generator spin, which produces electricity. The steam is then cooled and recirculated. The large cooling towers associated with nuclear plants are used to cool the steam after it has run through the turbines. A boiling water reactor works much the same way, except that the water surrounding the core boils and directly produces the steam, which is then piped to the turbine generator.

FIGURE 5.2

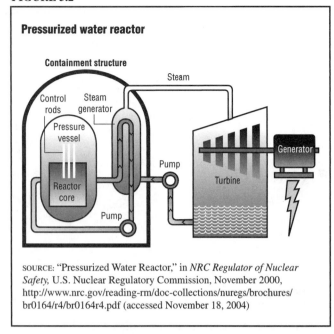

Pressurized water reactor

SOURCE: "Pressurized Water Reactor," in *NRC Regulator of Nuclear Safety,* U.S. Nuclear Regulatory Commission, November 2000, http://www.nrc.gov/reading-rm/doc-collections/nuregs/brochures/br0164/r4/br0164r4.pdf (accessed November 18, 2004)

FIGURE 5.3

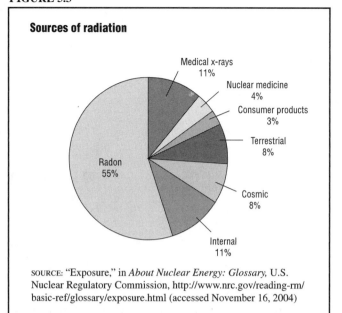

Sources of radiation

SOURCE: "Exposure," in *About Nuclear Energy: Glossary,* U.S. Nuclear Regulatory Commission, http://www.nrc.gov/reading-rm/basic-ref/glossary/exposure.html (accessed November 16, 2004)

The key problems in operating a nuclear power reactor include finding material (uranium 235) that will sustain a chain reaction, maintaining the reaction at a level that yields heat but does not escalate out of control and explode, and coping with the radiation produced by the chain reaction.

Radioactivity

Radioactivity is the spontaneous emission of energy and/or high-energy particles from the nucleus of an atom. One type of radioactivity is produced naturally and is emitted by radioactive isotopes (or radioisotopes), such as radioactive carbon (carbon 14) and radioactive hydrogen (H-3, or tritium). The energy and high-energy particles that radioactive isotopes emit include alpha rays, beta rays, and gamma rays.

Isotopes are atoms of an element that have the usual number of protons but different numbers of neutrons in their nuclei. For example, twelve protons and twelve neutrons comprise the nucleus of the element carbon. One isotope of carbon, C-14, has twelve protons and fourteen neutrons in its nucleus.

Radioisotopes (such as C-14) are unstable isotopes and their nuclei decay, or break apart, at a steady rate. Decaying radioisotopes produce other isotopes as they emit energy and/or high-energy particles. If the newly formed nuclei are radioactive as well, they emit radiation and change into other nuclei. The final products in this chain are stable, nonradioactive nuclei.

Radioisotopes reach our bodies daily, emitted from sources in outer space, and from rocks and soil on earth. Radioisotopes are also used in medicine and provide useful diagnostic tools. Figure 5.3 shows the sources of radiation.

As shown in Figure 5.3, radon is the largest source of radiation to which humans are exposed. This gas is formed in rocks and soil from the radioactive decay of radium. Most prevalent in the northern half of the United States, radon can enter cracks in basement walls and remain trapped there. Prolonged exposure to high levels of radioactive radon is thought to lead to lung cancer.

Decades after the discovery of radiation at the turn of the century by Antoine Henri Becquerel, Marie Curie, and Pierre Curie, other scientists determined that they could unleash energy by "artificially" breaking apart atomic nuclei. Such a process is called nuclear fission. Scientists learned that they could produce the most energy by bombarding the nuclei of an isotope of uranium called uranium 235 (U-235). The fission of U-235 releases several neutrons, which can then penetrate other U-235 nuclei. In this way the fission of a single U-235 atom could begin a cascading chain of nuclear reactions, as shown in Figure 5.4. If this series of reactions is regulated to occur slowly, as it is in nuclear power plants, the energy emitted can be captured for a variety of uses, such as generating electricity. If this series of reactions is allowed to occur all at once, as in a nuclear (atomic) bomb, the energy emitted is explosive. (Plutonium 239 can also be used to generate a chain reaction similar to that of U-235.)

Mining Nuclear Fuel

In the United States U-235, found in the form of ore, is used as nuclear fuel. The majority of uranium in the United States is found in two states: Wyoming and New Mexico. Ore that contains uranium is first located by geological methods such as drilling. Uranium-bearing ores are mined by methods similar to those used for other metal ores. However, uranium mining has the unique dan-

FIGURE 5.4

Nuclear chain reaction

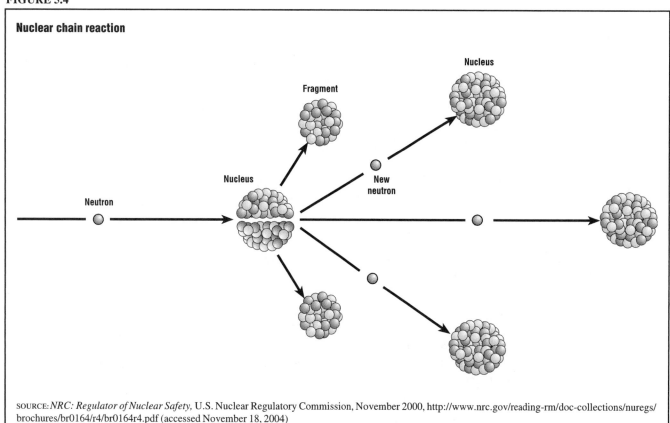

SOURCE: *NRC: Regulator of Nuclear Safety,* U.S. Nuclear Regulatory Commission, November 2000, http://www.nrc.gov/reading-rm/doc-collections/nuregs/brochures/br0164/r4/br0164r4.pdf (accessed November 18, 2004)

ger of exposure to radioactivity. Uranium atoms split by themselves at a slow rate, causing radioactive substances such as radon to accumulate slowly in the deposits.

After it is mined, uranium must be concentrated, because uranium ore generally contains only 0.1% uranium metal by weight. To concentrate the uranium, it goes through a process called milling. The nuclear fuel cycle, which is shown in Figure 5.5, begins with the milling of uranium ore. In milling the ore is first crushed, and then various chemicals are poured slowly through the crushed ore to dissolve out the uranium. The uranium is then precipitated from this chemical solution. The resulting material, called yellowcake because of its color, is 85% pure uranium by weight. But this uranium is 99.3% nonfissionable U-238 and only 0.7% fissionable U-235. Other processes result in enriched uranium, which is uranium that has a higher percentage of fissionable uranium than does yellowcake.

To produce enriched uranium, yellowcake is converted into uranium hexafluoride gas (UF6). This gas is loaded into cylinders, which are sent to a gaseous diffusion plant, shown in Figure 5.5, where uranium is made into reactor fuel (enriched). The enriched uranium is converted into oxide powder (UO2), which is made into fingertip-sized fuel pellets. The small pellets are less than one-half inch in diameter, but each one can produce as much energy as 120 gallons of oil. The pellets are stacked in tubes about twelve feet long, called rods. Many rods are bundled together in assemblies, and hundreds of these assemblies make up the core of a nuclear reactor.

DOMESTIC NUCLEAR ENERGY PRODUCTION

The percentage of U.S. electricity supplied by nuclear power grew considerably during the 1970s and 1980s before leveling off in the 1990s. In 1973 nuclear power supplied only 4.5% of the total electricity generated in the United States. By 2003 nuclear power's 19.8% share of electricity production was down only slightly from the 2001 all-time high of 20.6%. In 2002 nuclear power plant operations provided an all-time high of nuclear electricity net generation of 780.1 billion kilowatt-hours.

The increase in production of electricity in nuclear power plants was achieved largely through an increase in average capacity factor. The capacity factor is the proportion of electricity produced to what could have been produced at full-power operation. In 2002 the capacity factor was at an all-time high of 90.3%. In 2003 the capacity factor was 88.2%. Better training for operators, longer operating cycles between refueling, and control-system improvements contributed to increased plant performance.

In 2003 and 2004, 104 nuclear reactors were in operation (see Figure 5.6 and Figure 5.7) in thirty-one states. In 2003 these 104 reactors had a total net generation of 763.7 billion kilowatt-hours of energy. Most of these reactors

FIGURE 5.5

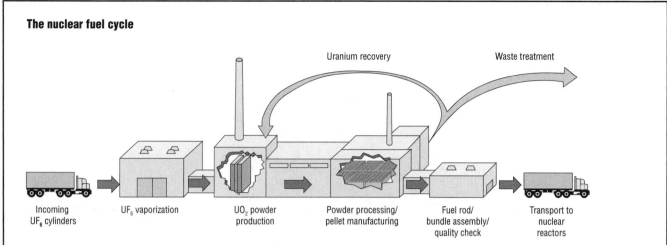

The nuclear fuel cycle

Uranium recovery

Waste treatment

Incoming UF₆ cylinders

UF₆ vaporization

UO₂ powder production

Powder processing/ pellet manufacturing

Fuel rod/ bundle assembly/ quality check

Transport to nuclear reactors

SOURCE: "The Nuclear Fuel Cycle," in *NRC Regulator of Nuclear Safety,* U.S. Nuclear Regulatory Commission, November 2000, http://www.nrc.gov/ reading-rm/doc-collections/nuregs/brochures/br0164/r4/br0164r4.pdf (accessed November 18, 2004)

are located east of the Mississippi River, where the demand for electricity is high. The number of plants operating is lower than the 135 reactors that were either planned, in construction, or in operation in 1974.

Since 1978 no new nuclear power plants have been ordered, and many have closed. (See Table 5.1.) Several factors have contributed to the slowdown in U.S. nuclear reactor construction. Overall costs have increased as a result of more expensive financing, partly influenced by longer delays for licensing. Expenses have also increased because of regulations that were instituted as a result of the Three Mile Island incident. Operating costs have been higher than expected as well. Originally, it was projected that plants could run almost 90% of the time, with brief pauses for refueling, but for many years this was not the case.

OUTLOOK FOR DOMESTIC NUCLEAR ENERGY

In its *Annual Energy Outlook 2004* (2004), the Energy Information Administration (EIA) predicted that the capacity of nuclear power plants will be maintained as they age through 2025. Thus, electricity generation by these plants is projected to increase slightly from 2002 to 2025, but its contribution to U.S. electricity production will decline as electricity generation by coal, natural gas, and renewables increases.

No nuclear units are projected to be retired between 2002 and 2025, nor are any new nuclear units expected to become operable during that time span because natural gas and coal-fired plants are projected to be more economical. However, the Nuclear Regulatory Commission (NRC) approved eighteen applications for power uprates for plants in 2002 and another nine were approved or pending as of 2003. (A power uprate is an increase in the power output of a nuclear power plant, which is accomplished by adding a more highly enriched uranium fuel to

FIGURE 5.6

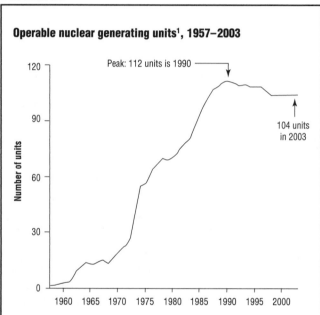

Operable nuclear generating units[1], 1957–2003

Peak: 112 units is 1990

104 units in 2003

Number of units

[1]Units holding full-power operating license, or equivalent permission to operate. Note: Data are at end of year.

SOURCE: Adapted from "Figure 9.1. Nuclear Generating Units: Operable Units, 1957–2003," in *Annual Energy Review 2003,* U.S. Department of Energy, Energy Information Administration, Office of Energy Markets and End Use, September 7, 2004, http://www.eia.doe.gov/emeu/aer/pdf/ aer.pdf (accessed September 28, 2004)

the current fuel. The plant must be able to operate safely at the higher power level.)

INTERNATIONAL PRODUCTION

According to the *Annual Energy Review 2003,* in 2002 nuclear power provided about 17% of the world's electricity and 6.6% of total energy. Although the United States is the largest producer of nuclear power, it trails

FIGURE 5.7

Operating nuclear reactors, June 2004

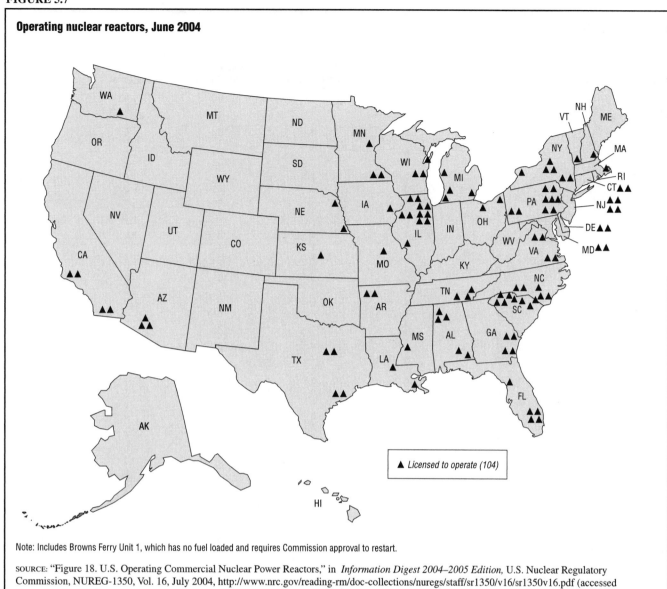

Note: Includes Browns Ferry Unit 1, which has no fuel loaded and requires Commission approval to restart.

SOURCE: "Figure 18. U.S. Operating Commercial Nuclear Power Reactors," in *Information Digest 2004–2005 Edition,* U.S. Nuclear Regulatory Commission, NUREG-1350, Vol. 16, July 2004, http://www.nrc.gov/reading-rm/doc-collections/nuregs/staff/sr1350/v16/sr1350v16.pdf (accessed November 16, 2004)

other Western countries in the proportion of national electrical production generated by nuclear power. That is primarily because oil, natural gas, and coal have been more accessible in the United States than in other countries.

In 2002 the United States led the world in nuclear power generation with 780.1 billion kilowatt-hours, followed by France's 414.9 billion and Japan's 295.1 billion. (See Table 5.2.) These three countries together generated 58% of the world's nuclear electric power. France had by far the highest proportion (approximately 78%) of its electrical power produced by nuclear energy, followed by Belgium (59%), Sweden (46%), and Switzerland (41%). Japan produces 28% of its electricity by nuclear power generation, and the United States only 20%.

In its *Information Digest 2004–2005,* the Nuclear Regulatory Commission reported that as of December 31, 2003, there were 438 nuclear power plants in operation worldwide and fifty-four were under construction, on order, or had their construction halted, including eight in India, eight in South Korea, six in Russia, five in the Ukraine, five in Japan, and four in Romania.

Worldwide, the EIA projected in its *International Energy Outlook 2004* that nuclear generating capacity will increase slightly overall, from 353 gigawatts in 2001 to 385 gigawatts in 2025. The increase will peak in 2015 with a generating capacity of 407 gigawatts. Totals for individual countries will increase for some and decrease for others. Some countries whose capacities are projected to increase from 2001 to 2025 are:

• Canada, from ten gigawatts in 2001, increasing to fifteen gigawatts in 2020, and decreasing to twelve gigawatts in 2025

TABLE 5.1

Nuclear generating units, 1953–2003

Year	Orders[1]	Cancelled orders[2]	Construction permits[3]	Low-power operating licenses[4]	Full-power operating licenses[5]	Shutdowns[6]	Operable units[7]
1953	1	0	0	0	0	0	0
1954	0	0	0	0	0	0	0
1955	3	0	1	0	0	0	0
1956	1	0	3	0	0	0	0
1957	2	0	1	1	1	0	1
1958	4	0	0	0	0	0	1
1959	4	0	3	1	1	0	2
1960	1	0	7	1	1	0	3
1961	0	0	0	0	0	0	3
1962	2	0	1	7	6	0	9
1963	4	0	1	3	2	0	11
1964	0	0	3	2	3	1	13
1965	7	0	1	0	0	0	13
1966	20	0	5	1	2	1	14
1967	29	0	14	3	3	2	15
1968	16	0	23	0	0	2	13
1969	9	0	7	4	4	0	17
1970	14	0	10	4	3	0	20
1971	21	0	4	5	2	0	22
1972	38	7	8	6	6	1	27
1973	42	0	14	12	15	0	42
1974	28	9	23	14	15	2	55
1975	4	13	9	3	2	0	57
1976	3	1	9	7	7	1	63
1977	4	10	15	4	4	0	67
1978	2	13	13	3	4	1	70
1979	0	6	2	0	0	1	69
1980	0	15	0	5	2	0	71
1981	0	9	0	3	4	0	75
1982	0	18	0	6	4	1	78
1983	0	6	0	3	3	0	81
1984	0	6	0	7	6	0	87
1985	0	2	0	7	9	0	96
1986	0	2	0	7	5	0	101
1987	0	0	0	6	8	2	107
1988	0	3	0	1	2	0	109
1989	0	0	0	3	4	2	111
1990	0	1	0	1	2	1	112
1991	0	0	0	0	0	1	111
1992	0	0	0	0	0	2	109
1993	0	0	0	1	1	0	110
1994	0	1	0	0	0	1	109
1995	0	2	0	1	0	0	109
1996	0	0	0	0	1	1	109
1997	0	0	0	0[8]	0[8]	2	107
1998	0	0	0	0	0	3	104
1999	0	0	0	0	0	0	104
2000	0	0	0	0	0	0	104
2001	0	0	0	0	0	0	104
2002	0	0	0	0	0	0	104
2003	0	0	0	0	0	0	104
Total	259	124	177	132	132	28	—

[1]Placement of an order by a utility or government agency for a nuclear steam supply system.

[2]Cancellation by utilities of ordered units. Includes WNP 1; the licensee intends to request that the construction permit be cancelled. Does not include three units (Bellefonte 1 and 2 and Watts Bar 2) where construction has been stopped indefinitely.

[3]Issuance by regulatory authority of a permit, or equivalent permission, to begin construction. Numbers reflect permits issued in a given year, not extant permits.

[4]Issuance by regulatory authority of license, or equivalent permission, to conduct testing but not to operate at full power.

[5]Issuance by regulatory authority of full-power operating license, or equivalent permission. Units generally did not begin immediate operation.

[6]Ceased operation permanently.

[7]Total of nuclear generating units holding full-power licenses, or equivalent permission to operate, at the end of the year. Although Browns Ferry 1 was shut down in 1985, the unit has remained fully licensed and thus has continued to be counted as operable during the shutdown; in May 2002, the Tennessee Valley Authority announced its intention to have the unit resume operation in 2007.

[8]Under new regulations beginning in 1997, the terms "Low-power operating licenses" and "Full-power operating licenses" are no longer applicable; while no new licenses have been granted under the new regulations, applications were made in 2003 for three "Early site permits."

— = Not applicable.

Web Page: For related information, see http://www.eia.doe.gov/fuelnuclear.html.

SOURCE: "Table 9.1. Nuclear Generating Units, 1953–2003," in *Annual Energy Review 2003*, U.S. Department of Energy, Energy Information Administration, Office of Energy Markets and End Use, September 7, 2004, http://www.eia.doe.gov/emeu/aer/pdf/aer.pdf (accessed September 28, 2004)

TABLE 5.2

World net nuclear electric power generation, 1980, 1990, and 2002

(Billion kilowatthours)

Region and country	Nuclear electric power			Total		
	1980	1990	2002ᴾ	1980	1990	2002ᴾ
North America	**287.0**	**648.9**	**860.3**	**2,721.6**	**3,623.9**	**4,592.7**
Canada	35.9	69.2	71.0	367.9	468.6	548.9
Mexico	0.0	2.8	9.3	63.6	116.6	203.6
United States	251.1	576.9	780.1	2,289.6	3,038.0	3,839.3
Other	0.0	0.0	0.0	0.5	0.7	0.9
Central and South America	**2.2**	**9.0**	**19.2**	**308.2**	**ᴿ497.0**	**789.7**
Argentina	2.2	7.0	5.4	41.8	48.3	81.4
Brazil	0.0	1.9	13.8	138.3	219.6	339.0
Paraguay	0.0	0.0	0.0	0.8	27.0	48.4
Venezuela	0.0	0.0	0.0	32.0	57.6	87.0
Other	0.0	0.0	0.0	95.3	ᴿ144.5	233.9
Western Europe	**219.2**	**707.5**	**881.7**	**1,844.5**	**2,351.7**	**2,918.4**
Belgium	11.9	40.6	45.0	50.8	66.5	76.6
Finland	6.6	18.3	21.2	38.7	51.8	71.6
France	63.4	298.4	414.9	250.8	397.6	528.6
Germany	55.6	145.1	156.8	469.9	526.0	548.3
Italy	2.1	0.0	0.0	176.4	202.1	261.6
Netherlands	3.9	3.3	3.7	62.9	67.6	90.6
Norway	0.0	0.0	0.0	82.9	120.4	125.9
Spain	5.2	51.6	59.9	109.2	143.9	229.0
Sweden	25.3	64.8	65.4	94.3	141.5	142.8
Switzerland	12.9	22.4	25.9	46.4	53.0	63.5
Turkey	0.0	0.0	0.0	23.3	55.2	123.3
United Kingdom	32.3	58.7	83.6	265.1	295.2	360.8
Other	0.0	4.4	5.3	173.8	230.8	295.9
Eastern Europe and former U.S.S.R	**83.2**	**251.3**	**297.1**	**1,604.1**	**1,976.6**	**1,619.9**
Czech Republic	—	—	17.8	—	—	71.8
Kazakhstan	—	—	0.0	—	—	55.4
Poland	0.0	0.0	0.0	114.7	128.5	133.8
Romania	0.0	0.0	5.1	63.9	60.6	53.6
Russia	—	—	134.1	—	—	850.6
Ukraine	—	—	73.4	—	—	167.3
Other	83.2	251.3	66.7	1,425.6	1,787.5	287.4
Middle East	**0.0**	**0.0**	**0.0**	**92.4**	**228.9**	**490.0**
Iran	0.0	0.0	0.0	21.3	55.9	129.0
Saudi Arabia	0.0	0.0	0.0	20.5	64.9	138.2
Other	0.0	0.0	0.0	50.7	108.1	222.8
Africa	**0.0**	**8.4**	**12.0**	**189.2**	**307.5**	**453.9**
Egypt	0.0	0.0	0.0	18.3	41.4	81.3
South Africa	0.0	8.4	12.0	93.1	156.0	202.6
Other	0.0	0.0	0.0	77.8	110.1	170.0
Asia and Oceania	**92.7**	**279.9**	**489.3**	**1,280.5**	**ᴿ2,354.7**	**4,425.9**
Australia	0.0	0.0	0.0	87.7	146.4	210.3
China	0.0	0.0	23.5	285.5	590.3	1,575.1
India	3.0	5.6	17.8	119.3	275.5	547.2
Indonesia	0.0	0.0	0.0	13.5	46.5	99.3
Japan	78.6	192.2	295.1	549.1	822.1	1,044.0
South Korea	3.3	50.2	113.1	34.6	100.4	287.6
Taiwan	7.8	31.6	38.0	42.0	83.3	158.5
Thailand	0.0	0.0	0.0	13.6	43.7	102.4
Other	(s)	0.4	1.8	135.3	ᴿ246.4	401.4
World	**684.4**	**1,905.0**	**2,559.6**	**8,040.5**	**ᴿ11,340.2**	**15,290.5**

R=Revised.
P=Preliminary.
— = Not applicable.
(s)=Less than 0.05 billion kilowatthours.
Note: Totals may not equal sum of components due to independent rounding.
Web Page: For related information, see http://www.eia.doe.gov/international.

SOURCE: "Table 11.16. World Net Generation of Electricity by Type, 1980, 1990, and 2002 (Billion Kilowatthours)," in *Annual Energy Review 2003,* U.S. Department of Energy, Energy Information Administration, Office of Energy Markets and End Use, September 7, 2004, http://www.eia.doe.gov/emeu/aer/pdf/aer.pdf (accessed September 28, 2004)

- Japan, from forty-three gigawatts to fifty-seven gigawatts in 2020, and down to fifty-four gigawatts in 2025

- China, from two gigawatts in 2001 to twenty-one gigawatts in 2025

- India, from three to nine gigawatts

- United States, from ninety-eight to 103 gigawatts

- France, from sixty-three to seventy gigawatts

Some countries whose capacities are projected to decrease from 2001 to 2025 are:

- Germany, from twenty-one to zero gigawatts

- The United Kingdom, from twelve to four gigawatts

Some countries are planning to cut capacity radically or phase out nuclear power altogether. Sweden has legislated a nuclear phase-out by 2010 (although it is not clear that this is an attainable goal, since nearly half of Sweden's electricity comes from nuclear power). The Netherlands and Germany have strong conservation policies and well-organized antinuclear movements. The Netherlands is projected to phase out nuclear power by 2020 and Germany by 2025.

International Agreement on Safety

In September 1994 forty nations, including the United States, signed the International Convention on Nuclear Safety, an agreement that requires them to shut down nuclear power plants if necessary safety measures cannot be guaranteed. The agreement applies to land-based civil nuclear power plants and seeks to avert accidents like the 1986 explosion at Chernobyl in the former Soviet Union, the world's worst nuclear disaster to date. Ukraine, which inherited the Chernobyl plant after the collapse of the Soviet Union, signed the agreement. Signers must submit reports on atomic installations and, if necessary, make improvements to upgrade safety at the sites. Neighboring countries may call for an urgent study if they are concerned about a reactor's safety and the potential fallout that could affect their own population or crops.

POWER PLANT AGING

Like other types of power plants, nuclear power plants have a lifespan after which they must be retired because the cost of running them becomes too high. The generation of nuclear power plants currently operating was designed to last about thirty years. Some plants showed serious levels of deterioration after as few as fifteen years. However, some have lasted well beyond thirty-year expectations. Whenever a power plant—whether coal, natural gas, or nuclear—is retired, it must be demolished and cleaned up so that the site can be used again.

Decommissioning Nuclear Power Plants

Eventually, the more than 400 nuclear plants now operating worldwide will need to be retired. There are three methods of retiring, or decommissioning, a reactor: safe enclosure (mothballing), entombment, and immediate dismantling.

Safe enclosure involves removing the fuel from the plant, monitoring any radioactive contamination (which is usually very low or nonexistent), and guarding the structure to prevent anyone from entering until eventual dismantling and decontamination activities occur. Entombment, which was used at Chernobyl, involves permanently encasing the structure in a long-lived material such as concrete. This procedure allows the radioactive material to remain safely on-site but at one location. Immediate dismantling involves decontaminating and tearing down the facility within a few months or years. This process is initially more expensive than the other options but removes the long-term costs of monitoring both the structure and the radiation levels. It also frees the site for other uses, including even the construction of another nuclear power plant. Nuclear power officials are also seeking alternative uses for nuclear shells, including the conversion of old nuclear plants to gas-fired plants.

Paying for the closing, decontamination, or dismantling of nuclear plants has become an issue of intense public debate. The early closing of several plants owing to safety concerns and poor economic performance has raised the question of how the costs of decommissioning can be covered. Industry reports contend that the cost of decommissioning retired plants and handling radioactive wastes will continue to escalate, causing serious financial problems for electric utilities. In addition to coal plants, new technologies, such as wind turbines and geothermal energy, are less expensive than nuclear energy plants when all factors are considered.

A NEW GENERATION OF NUCLEAR PLANTS?

In January 2000 the United States entered into a collaboration with international partners on nuclear energy. Formally chartered in 2001, the Generation IV International Forum (GIF) is a group of leading nuclear nations that agree that nuclear energy is important to future world energy security and economic prosperity. These countries are dedicated to joint development of the "next generation" of nuclear energy systems. Members of GIF are the United States, Argentina, Brazil, Canada, France, Japan, the Republic of Korea, South Africa, Switzerland, and the United Kingdom.

The GIF countries have selected six nuclear power technologies to develop for the future. These "Generation IV" nuclear energy systems would follow the three other periods of nuclear reactor development: (a) Generation I experimental reactors developed in the 1950s and 1960s; (b) Generation II large, central-station nuclear power reactors, such as the 104 plants still operating in the United States, built in the 1970s and 1980s; and (c) Generation

III advanced light-water reactors built in the 1990s, primarily in East Asia, to meet that region's expanding electricity needs.

In 2003 GIF examined more than one hundred reactor concepts and identified nineteen as potentially viable. One new concept, the advanced high-temperature reactor (AHTR), may reduce electricity production costs and may be able to produce hydrogen. GIF research and development began in 2004.

NUCLEAR SAFETY PROBLEMS

Problems have plagued the nuclear industry almost from the beginning. Some plant-site selections have been considered questionable, especially those near earthquake fault lines. Several major accidents have occurred worldwide since the late 1970s.

In the United States

THREE MILE ISLAND. On March 28, 1979, the Three Mile Island nuclear facility near Harrisburg, Pennsylvania, was the site of the worst nuclear accident in American history. Information released several years after the accident revealed that the plant came much closer to meltdown than either the NRC or the industry had previously indicated. Temperatures inside the reactor were first said to have been 3,500 degrees Fahrenheit but are now known to have reached at least 4,800 degrees. The temperature needed to melt uranium dioxide fuel is 5,080 degrees Fahrenheit. When meltdown occurs, an uncontrolled explosion may result, unlike the controlled nuclear reaction of normal operation.

The emergency core cooling system at Three Mile Island was designed to dump water on the hot core and spray water into the reactor building to stop the production of steam. During the accident, however, the valves leading to the emergency water pumps closed. Another valve was stuck in the open position, drawing water away from the core, which then became partially uncovered and began to melt. The emergency core cooling system then began drawing water out of the basement supply and reusing it, contaminating the reactor pump, although limiting the radiation contamination to the interior of the building.

Although safety systems at the Three Mile Island facility performed well and the radiation leak was relatively small, the Three Mile incident was sensationalized in the media. Antinuclear proponents urged residents to evacuate, adding to stress levels and panic. In the end this nuclear accident resulted in no deaths or injuries to plant workers or members of the nearby community. On average, area residents were exposed to less radiation than that of a chest X-ray. Nevertheless, the Three Mile Island incident raised concerns about nuclear safety, which resulted in safety enhancements in the nuclear power industry and at the NRC. Antinuclear sentiment was fueled as well, heightening Americans' criticism of nuclear power as an energy source.

The nuclear reactor in Unit 2 at Three Mile Island has been in monitored storage since it underwent cleanup. Operation of the reactor in Unit 1 resumed in 1985.

PEACH BOTTOM. In 1987 the NRC shut down the Peach Bottom nuclear plant in Delta, Pennsylvania, because control-room operators were found sleeping on duty. This was the first plant shut-down solely because of operator violations and misconduct.

In the Former Soviet Union

CHERNOBYL. On April 26, 1986, the most serious nuclear accident ever occurred at Chernobyl, a four-reactor nuclear plant complex located in the former Soviet Union (now Ukraine) near Kiev. At least thirty-one people died and hundreds were injured when one of the four reactors exploded during a badly-run test of the reactor. Everyone in the vicinity was exposed to radiation, and radioactive particles were released into the atmosphere. About 500 people were hospitalized, and medical experts estimate that 6,000 to 24,000 cancer-related deaths have occurred over the intervening years as a result of the released radiation.

The cleanup was one of the biggest projects of its kind ever undertaken. Helicopters dropped 5,000 tons of limestone, sand, clay, lead, and boron on the smoldering reactor to stop the radiation leakage and reduce the heat. Workers built a giant steel and cement sarcophagus (stone coffin) to enshrine the remains of the reactor and contain the radioactive waste. Approximately 135,000 people were evacuated from a 300-square-mile area around the power station. Topsoil had to be removed in a nineteen-mile area and buried as nuclear waste. Buildings were washed down and the newly contaminated water and soil were carted away and buried. Agricultural products from areas nearby were declared unmarketable throughout Europe. Many of the 600,000 people involved in the immediate cleanup have suffered long-term effects of radiation exposure.

Some people at Chernobyl received 400 rems of radiation immediately following the explosion. (A rem is a standard measure of the whole-body dose of radiation. Under normal conditions, a person receives about one-tenth of a rem annually.) With a dose of twenty-five rems, a person's blood begins to change. The DNA is damaged, preventing the bone marrow from producing red and white blood cells. Sickness starts at 100 rems and severe sickness at 200 rems. Death of half the population occurs at 400 rems, and death within a week can be expected for anyone exposed to 600 rems. One estimate is that 17.5 million people, including 2.5 million children under seven years old, have had some significant exposure to radiation from Chernobyl.

In 1995 the United Nations reported that the prevalence of illnesses of all kinds was 30% above normal in the Ukraine; incidences of depression, alcoholism, and divorce were on the rise. Although the Chernobyl reactor is in the Ukraine, the neighboring nation of Belarus (with a population of 10 million people) suffered more human and ecological damage from fallout because of the prevailing winds. Thyroid cancers were 285 times pre-Chernobyl levels in Belarus in 1995, especially among children. Belarus's cabinet minister claimed at that time that a quarter of his country's national income was being spent on alleviating the effects of the disaster. As many as 375,000 people in Belarus, Ukraine, and Russia were displaced because of the accident. Contaminated forests spread radioactivity through fires, and seepage from the reactor has polluted waterways as far away as the Black Sea, some 350 miles away.

In April 1996 a fire in the woods near the Chernobyl reactor once more alarmed observers, who feared further danger to the entombed reactor and the release of radioactivity from soils and vegetation in the area. In addition, the sarcophagus built to contain the damaged reactor is reported to be crumbling. However, the Ukraine, which suffers from 40% unemployment and other enormous economic woes, claims it needs the power from the remaining reactor and that completely shutting down Chernobyl would cost too much.

Some high-technology companies have developed robots to clean up hazardous sites such as Chernobyl. Such efforts would, however, produce yet another problem: what to do with the radioactive waste once the Chernobyl sarcophagus is entered and cleaned up.

TYPES OF RADIOACTIVE WASTE

Working in a laboratory in Chicago, Illinois, in 1942, Italian physicist Enrico Fermi assembled enough uranium to cause a nuclear fission reaction. His discovery transformed both warfare and energy production, but the experiment also produced a small packet of radioactive waste materials that will remain dangerous for hundreds of thousands of years. That "first" radioactive waste lies buried under a foot of concrete and two feet of dirt on a hillside in Illinois.

Radioactive waste material is produced at all stages of the nuclear fuel cycle, from the initial mining of the uranium to the final disposal of the spent fuel from the reactor. (See Figure 5.5.) "Radioactive waste" is a term that encompasses a broad range of material with widely varying characteristics. Some is barely radioactive and safe to handle, while other types are intensely hot in terms of both temperature and radioactivity. Some waste decays to safe levels of radioactivity in a matter of days or weeks, while other types will remain dangerous for thousands of years. The U.S. Department of Energy (DOE) and NRC define the major types of radioactive waste.

Uranium Mill Tailings

Uranium mill tailings are sand-like wastes produced in uranium refining operations. Although they emit low levels of radiation, their large volumes (10–15 million tons annually) pose a hazard, particularly from radon emissions and groundwater contamination. Since it was not until the early 1970s that the dangers of uranium mill tailings were realized, many miners and residents in the western United States had unsafe exposure to them. Cancer incidences are high among miners who worked prior to the 1970s.

Low-level Waste

Low-level waste, which contains varying lesser levels of radioactivity, includes trash (such as wiping rags, swabs, and syringes), contaminated clothing (such as shoe covers and protective gloves), and hardware (such as luminous dials, filters, and tools). This waste comes from nuclear reactors, industrial users, government users (but not nuclear weapons sites), research universities, and medical facilities. In general, low-level waste decays relatively quickly (in ten to one hundred years).

High-level Waste

Spent nuclear fuel (used reactor fuel) is high-level radioactive waste. Uranium fuel can be used for twelve to eighteen months, and then it is no longer as efficient in splitting its atoms and producing heat, which ultimately is used to generate electricity. Therefore, it must be removed from the reactor and replaced with fresh fuel. Some of the spent fuel is reprocessed to recover the usable uranium and plutonium, but the radioactive material that remains is dangerous for thousands of years.

Transuranic (TRU) Wastes

Transuranic (TRU) wastes are eleven human-made radioactive elements with atomic numbers greater than that of uranium (ninety-two) and therefore beyond ("trans-") uranium ("-uranic") on the periodic chart of the elements. Their half-lives—the time it takes for half the radioisotopes present in a sample to decay to nonradioactive elements—are thousands of years. They are found in trash produced mainly by nuclear weapons plants and are therefore part of the nuclear waste problem but not directly the concern of nuclear power utilities.

Mixed Waste

Mixed waste is high-level, low-level, or TRU waste that also contains hazardous nonradioactive waste. Such waste poses serious institutional problems, because the radioactive portion is regulated by the DOE or NRC under the Atomic Energy Act (AEA; PL 83–703), while the Environmental Protection Agency regulates the nonradioactive elements under the Resource Conservation and Recovery Act (RCRA; PL 95–510).

RADIOACTIVE WASTE DISPOSAL

Disposing of radioactive waste is unquestionably one of the major problems associated with the development of nuclear power; radioactive waste is also a by-product of nuclear weapons plants, hospitals, and scientific research. Although federal policy is based on the assumption that radioactive waste can be disposed of safely, new storage and disposal facilities for all types of radioactive waste have frequently been delayed or blocked by concerns about safety, health, and the environment.

The highly toxic wastes must be isolated from the environment until the radioactivity decays to a safe level. In the case of plutonium, for example, the half-life is 26,000 years. At that rate, it will take at least 100,000 years before radioactive plutonium is no longer dangerous. Any facilities built to store such materials must last at least that long.

Regulation of Radioactive Waste Disposal

The Nuclear Regulatory Commission, other federal agencies, and the states regulate radioactive materials. As of 2004, the NRC had entered into agreements regarding radioactive waste with 33 states, which are called Agreement States. These states, under their agreements with the NRC, regulate the management, storage, and disposal of certain nuclear waste within their states.

Disposal of Uranium Mill Tailings

Mill tailings are usually deposited in large piles next to the mill that processed the ore. In 1978 Congress passed the Uranium Mill Tailing Radiation Control Act (PL95–604). This law requires mill owners to follow Environmental Protection Agency (EPA) standards for cleanup of uranium and thorium after milling operations have permanently closed. They must cover the mill tailings to control the release of radon gas; the cover must effectively control radon releases for 1,000 years. Figure 5.8 shows the locations of uranium mill tailings disposal sites. The sites are located in the western United States because deposits of uranium ore are more plentiful there.

Disposal of Low-level Waste

Low-level waste is classified according to its potential hazards. Commercial sites that dispose of these wastes must be licensed by either the NRC or Agreement States. In 2004 three low-level waste facilities were operational in the United States. These facilities accept a broad range of low-level radioactive waste and are located in Barnwell, South Carolina, and Richland, Washington. In addition, Envirocare of Utah accepts large amounts of mill tailings and low-level waste, such as contaminated soil or debris from demolished buildings and from facilities that have shut down. Four low-level radioactive waste facilities have been closed: West Valley, New York (closed 1975); More-

FIGURE 5.8

Uranium mill tailings disposal sites

SOURCE: "Locations of Uranium Mill Tailings Sites," in *Radioactive Waste: Production, Storage, Disposal,* U.S. Nuclear Regulatory Commission, Office of Public Affairs, May 2002, http://www.nrc.gov/reading-rm/doccollections/nuregs/brochures/br0216/r2/br0216r2.pdf (accessed November 31, 2004)

head/Maxey Flats, Kentucky (closed 1977); Sheffield, Illinois (closed 1978); and Beatty, Nevada (closed 1993).

The design of a low-level waste facility is shown in Figure 5.9. Wastes are buried in shallow underground sites in the specially designed canisters in which they are shipped as shown in the diagram. Underground storage may or may not include protection by concrete vaults.

The Low-level Radioactive Waste Policy Amendments Act of 1985 (PL 99–240) encouraged states to enter into compacts, which are legal agreements among states for low-level radioactive waste disposal. Figure 5.10 shows these compacts. Although each compact is responsible for the development of disposal capacity for the low-level waste generated within the compact, new disposal sites have yet to be built. As Figure 5.10 shows, two of the three operational low-level sites are located in the Northwest Compact and one is in the Southeast Compact. Facilities located in compacts with no current low-level waste disposal sites must petition the compact to export their low-level radioactive waste to one of the three operating disposal sites.

Disposal of High-level Waste

A major step toward shifting the responsibility for disposal of high-level radioactive wastes (spent fuel) from the nuclear power industry to the federal government was taken in 1982, when Congress passed the Nuclear Waste Policy Act (PL 97–425). It provided the first comprehensive national policy and detailed timetable for the manage-

FIGURE 5.9

Low-level waste disposal site

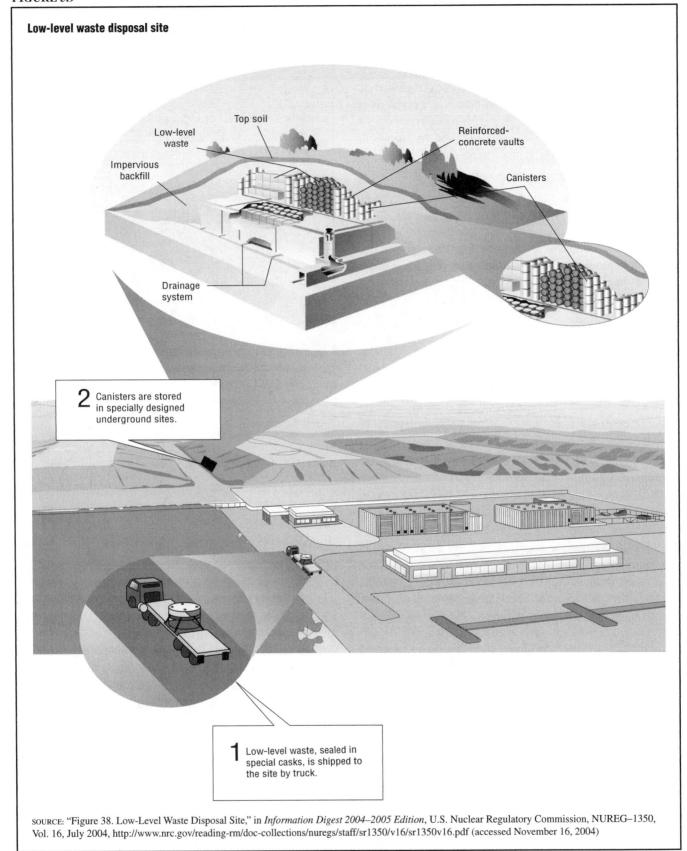

SOURCE: "Figure 38. Low-Level Waste Disposal Site," in *Information Digest 2004–2005 Edition*, U.S. Nuclear Regulatory Commission, NUREG–1350, Vol. 16, July 2004, http://www.nrc.gov/reading-rm/doc-collections/nuregs/staff/sr1350/v16/sr1350v16.pdf (accessed November 16, 2004)

ment and disposal of high-level nuclear waste and authorized construction of the first high-level nuclear waste repository. A 1987 amendment to the Nuclear Waste Poli-

cy Act directed investigation of Yucca Mountain in Nevada as a potential site. In 2004 no long-term, high-level permanent waste disposal repository for spent fuel existed. While

FIGURE 5.10

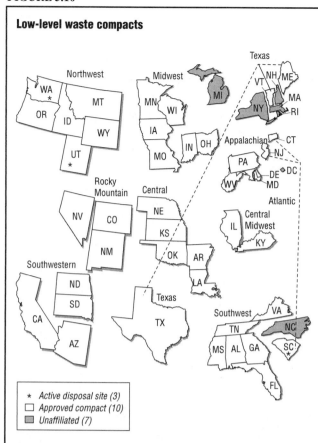

Low-level waste compacts

★ Active disposal site (3)
□ Approved compact (10)
▨ Unaffiliated (7)

Note: Data as of March 2004.
Alaska and Hawaii belong to the Northwest compact.
Puerto Rico is unaffiliated.
There are three active, licensed low-level waste disposal facilities located in agreement states.
Barnwell, located in Barnwell, South Carolina– Currently, Barnwell accepts waste from all U.S. generators except those in Rocky Mountain and Northwest compacts. Beginning in 2008, Barnwell will only accept waste from the Atlantic compact states (Connecticut, New Jersey, and South Carolina). Barnwell is licensed by the State of South Carolina to receive waste in Classes A–C.
Hanford, located in Hanford, Washington– Hanford accepts waste from the Northwest and Rocky Mountain compacts. Hanford is licensed by the state of Washington to receive waste in classes A–C.
Envirocare, located in Clive, Utah – Envirocare accepts waste from all regions of the United States. Envirocare is licensed by the state of Utah for class A waste only.

SOURCE: "Figure 39. U.S. Low-Level Waste Compacts," in *Information Digest 2004–2005 Edition,* U.S. Nuclear Regulatory Commission, NUREG-1350, Vol. 16, July 2004, http://www.nrc.gov/reading-rm/doc-collections/nuregs/staff/sr1350/v16/sr1350v16.pdf (accessed November 16, 2004)

waiting for the development of the Yucca Mountain site, spent fuel was being stored at away-from-reactor storage facilities, such as the General Electric Company facility in Morris, Illinois, or at the nuclear power plants that generated the waste. (See Figure 5.7 for the locations of operating nuclear reactors in the United States.)

On July 9, 2002, the U.S. Senate joined the U.S. House of Representatives in approving the Yucca Mountain site. The State of Nevada, however, challenged the constitutionality of this resolution. Nevertheless, in 2004, federal courts dismissed all challenges to the site selection of Yucca Mountain but did call into question the legality

of the 10,000-year standard for the facility set by the Environmental Protection Agency because the EPA did not accept the National Academy of Sciences recommendation of a higher standard of perhaps 300,000 years. While the agency works on a new standard, Congress could decide to allow the old one to stay in place. Congress approved $2 million for the state's work on the project and $8 million for local governments for fiscal year 2005, an increase from the state's $1 million and local government's $4 million received for fiscal year 2004. The DOE must still submit its license application for the proposed Yucca Mountain nuclear repository. It will take about five years to construct the facility, which is already more than a decade behind schedule.

Figure 5.11 shows the location of the proposed radioactive waste disposal facility at Yucca Mountain. The proposed repository would look like a large mining complex. It would have facilities on the surface for handling and packaging nuclear waste and a large mine about 1,000 feet underground. Here, plans call for the waste to be placed in sealed metal canisters arranged vertically in the floor of underground tunnels. Above-ground facilities would cover approximately 400 acres and be surrounded by a three-mile buffer zone. Underground, about 1,400 acres would be mined, consisting of tunnels leading to the areas where waste containers would be placed and service areas near the shafts and ramps that provide access from the surface. This type of "deep geologic" disposal is widely considered by governments, scientists, and engineers to be the best option for isolating highly radioactive waste.

The repository would be designed to contain radioactive material by using layers of human-made and natural barriers. Regulations require that a repository isolate waste until the radiation decays to a level that is about the same as that from a natural underground uranium deposit. This decay time was originally estimated at be about 10,000 years, but the National Academy of Sciences has more recently recommended a higher standard of about 300,000 years.

After the repository has been filled to capacity, regulations require the DOE to keep the facility open and to monitor it for at least fifty years from the fill date. This will allow experts to monitor conditions inside the repository and retrieve spent fuel if necessary. Eventually, the repository shafts will be filled with rock and earth and sealed. At the ground level, facilities will be removed, and, as much as possible, the DOE will take steps to return the site to its original condition.

Scientists assume that over thousands of years, some of the human-made barriers in a repository will break down. Once that happens, natural barriers will be counted on to stop or slow the movement of radiation particles. The most

FIGURE 5.11

Yucca Mountain waste site

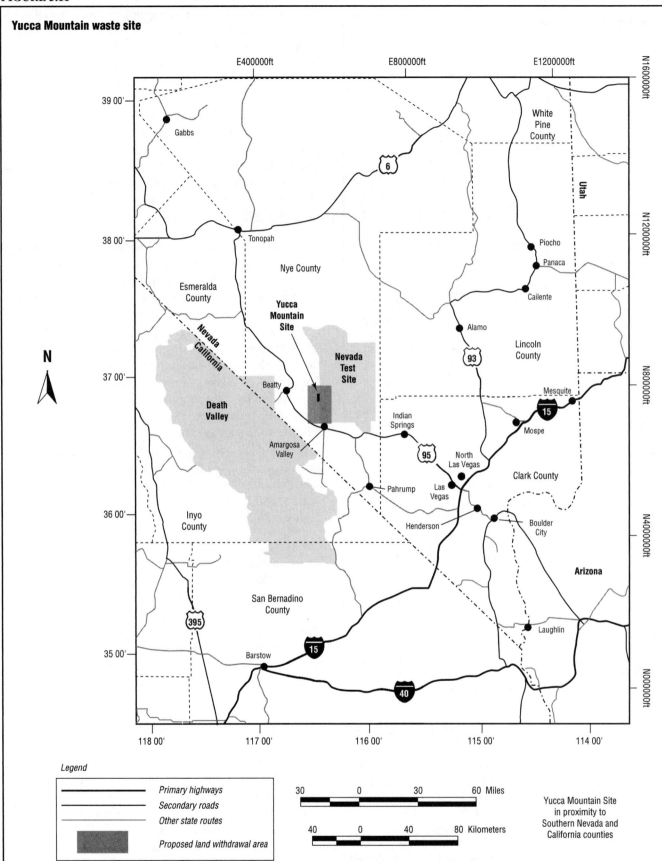

Legend

Primary highways

Secondary roads

Other state routes

Proposed land withdrawal area

30   0   30   60 Miles

40   0   40   80 Kilometers

Yucca Mountain Site
in proximity to
Southern Nevada and
California counties

SOURCE: "Figure 1-5. Map Showing the Location of Yucca Mountain in Relation to Major Highways; Surrounding Counties, Cities, and Towns in Nevada and California; the Nevada Test Site; and Death Valley National Park," in *Yucca Mountain Science and Engineering Report: Technical Information Supporting Site Recommendation Consideration Revision 1,* U.S. Department of Energy, February 2002, http://www.ymp.gov/documents/ser_b/ (accessed November 18, 2004)

TABLE 5.3

Total commercial spent nuclear fuel discharges, 1968–2002

Reactor type	Number of assemblies		
	Stored at reactor sites	Stored at away-from-reactor facilities	Total
Boiling-water reactor	90,398	2,957	93,355
Pressurized-water reactor	69,800	491	70,291
High-temperature gas cooled reactor	1,464	744	2,208
Total	**161,662**	**4,192**	**165,854**
	Metric tonnes of uranium (MTU)		
Boiling-water reactor	16,153.60	554	16,707.60
Pressurized-water reactor	30,099.00	192.6	30,291.60
High-temperature gas cooled reactor	15.4	8.8	24.2
Total	**46,268.00**	**755.4**	**47,023.40**

MTU=Metric tonnes of uranium.
Notes: A number of assemblies discharged prior to 1972, which were reprocessed, are not included in this table (no data is available for assemblies reprocessed before 1972). Totals may not equal sum of components because of independent rounding.

SOURCE: "Table 1. Total U.S. Commercial Spent Nuclear Fuel Discharges, 1968–2002," in *Energy Information Administration Spent Nuclear Fuel Data, Detailed United States as of December 31, 2002,* U.S. Department of Energy, Energy Information Administration, October 1, 2004, http://www.eia.doe.gov/cneafr/nuclear/spent_fuel/ussnfdata.html (accessed November 18, 2004)

likely way for particles to reach humans and the environment would be through water, which is why the low water tables at Yucca Mountain are so crucial. Yucca Mountain also has certain chemical properties that act as another barrier to the movement of radioactive particles. Minerals in the rock called zeolites would stick to the particles and slow their movement throughout the environment.

The long delay in providing disposal sites for spent nuclear fuel, coupled with the accelerated pace at which nuclear plants are being retired, has created a crisis in the industry. Several aging plants are being maintained at a cost of $20 million a year for each reactor simply because there is no place to send the waste once the plants are decommissioned. Table 5.3 shows that as of 2002 more than 47,000 metric tons of nuclear uranium waste were sitting in spent fuel pools at the 104 operating and nineteen permanently closed nuclear power plants. When a nuclear plant shuts down, the nuclear waste and the radioactive equipment stay on the premises because there is no place to put them. As a result, every nuclear power plant in the United States has become a temporary nuclear waste disposal site. Plants that close must wait until a repository opens to be decommissioned or dismantled.

Disposal of Transuranic Waste

The first disposal facility licensed to dispose of transuranic waste opened on March 26, 1999. The Waste Isolation Pilot Plant (WIPP) is located in the desert in southeastern New Mexico. WIPP facilities include disposal rooms mined 2,150 feet underground in a 2,000-foot-thick salt formation that has been stable for more than 200 million years.

Disposal of Mixed Waste

As mentioned previously, mixed waste is jointly regulated by the Environmental Protection Agency, which regulates the hazardous waste component of mixed waste, and the Nuclear Regulatory Commission and the Department of Energy, which regulate the radioactive component of mixed waste. There are a variety of commercial facilities in the United States, including Envirocare (see "Disposal of Low-level Waste"), that accept mixed waste.

CHAPTER 6
RENEWABLE ENERGY

RENEWABLE ENERGY DEFINED

Imagine energy sources that use no oil, produce no pollution, cannot be affected by political events and cartels, create no radioactive waste, and yet are economical. Although it sounds impossible, some experts claim that technological advances could make wide use of renewable energy sources possible within a few decades. Renewable energy is energy that is naturally regenerated and is, therefore, virtually unlimited. Sources include the sun (solar), wind, water (hydropower), vegetation (biomass), and the heat of the earth (geothermal).

Solar energy, wind energy, hydropower, and geothermal power are all renewable, inexpensive, and clean sources of energy. Each of these alternative energy sources has advantages and disadvantages, and many observers hope that one or more of them may eventually provide a substantially better energy source than conventional fossil fuels. As the United States and the rest of the world continue to expand their energy needs—putting a strain on the environment and nonrenewable resources—alternative sources of energy continue to be explored.

A HISTORICAL PERSPECTIVE

Before the eighteenth century, most energy came from renewable sources. People burned wood for heat, used sails to harness the wind and propel boats, and installed water wheels on streams to grind grain. The large-scale shift to nonrenewable energy sources began in the 1700s with the Industrial Revolution, a period marked by the rise of factories, first in Europe and then in North America. As demand for energy grew, coal replaced wood as the main fuel. Coal was the most efficient fuel for the steam engine, one of the most important inventions of the Industrial Revolution.

Until the early 1970s most Americans were unconcerned about the sources of the nation's energy. Supplies of coal and oil, which together provided more than 90% of U.S. energy, were believed to be plentiful. The decades preceding the 1970s were characterized by cheap gasoline and little public discussion of energy conservation.

That carefree approach to energy consumption ended in the 1970s. A fuel oil crisis made Americans more aware of the importance of developing alternative sources of energy to supplement and perhaps eventually even replace fossil fuels. In major cities throughout the United States, gasoline rationing became commonplace, lower heat settings for offices and living quarters were encouraged, and people waited in line to fill their gas tanks. In a country where mobility and personal transportation were highly valued, the oil crisis was a shocking reality for many Americans. As a result, the administration of President Jimmy Carter encouraged federal funding for research into alternative energy sources.

In 1978 the U.S. Congress passed the Public Utilities Regulatory Policies Act (PURPA; PL 95-617), which was designed to help the struggling alternative energy industry. The act exempted small alternative producers from state and federal utility regulations and required existing local utilities to buy electricity from them. The renewable energy industries responded by growing rapidly, gaining experience, improving technologies, and lowering costs. This act was the single most important factor in the development of the commercial renewable energy market.

In the 1980s President Ronald Reagan decided that private-sector financing for the short-term development of alternative energy sources was better than public-sector financing. As a result, he proposed the reduction or elimination of federal expenditures for alternative energy sources. Although funds were severely cut, the U.S. Department of Energy (DOE) continued to support some research and development to explore alternate sources of energy. President Bill Clinton's administration reemphasized the importance of renewable energy and increased funding in several areas. The George W. Bush administration believed that renewable and alternative fuels offer hope for the future but also considered that only a small portion of America's energy needs as of the early 2000s were offset by renewables. The Bush administration supported funding for research and development in renewable technologies and tax credits for the purchase of hybrid and fuel cell cars.

DOMESTIC RENEWABLE ENERGY USAGE

Renewable energy contributes only a small portion of the nation's energy supply. In 2003 the United States consumed approximately 6.2 quadrillion Btu of renewable energy, about 6% of the nation's total energy consumption. (See Table 6.1.) Biomass sources (wood, waste, and alcohol) contributed 2.9 quadrillion Btu, while hydroelectric power provided 2.8 quadrillion Btu. Together, biomass and hydroelectric power provided 92% of renewable energy in 2003, or around 6% of all energy as shown in Figure 6.1. Geothermal energy was the third largest source, with about 0.3 quadrillion Btu. Solar power contributed 0.06 quadrillion Btu, and wind provided 0.1 quadrillion Btu.

BIOMASS ENERGY

Biomass refers to organic material such as plant and animal waste, wood, seaweed and algae, and garbage. The use of biomass is not without environmental problems. Deforestation can occur from widespread wood use if forests are clear-cut, resulting in the possibility of soil erosion and mudslides. Burning wood, like burning fossil fuels, also pollutes the environment. Biomass can be burned directly or converted to biofuel by thermochemical conversion and biochemical conversion.

Direct Burning

Direct combustion is the easiest and most commonly used method of using biomass as fuel. Materials such as dry wood or agricultural wastes are chopped and burned to produce steam, electricity, or heat for industries, utilities, and homes. Industrial-size wood boilers are operating throughout the country, and the Department of Energy (DOE) maintains that many more will be built during the next decade. The burning of agricultural wastes is also becoming more widespread. In Florida, sugarcane producers use the residue from the cane to generate much of their energy.

Wood burning in stoves and fireplaces is another example of direct burning of biomass for energy, in this case heat. In the United States, residential use of wood as fuel generated 359 trillion Btu in 2003. In comparison, the generation of Btu from burning wood in the home in the 1980s was about 850–950 trillion Btu according to the Energy Information Administration's *Annual Energy Review 2003*.

FIGURE 6.1

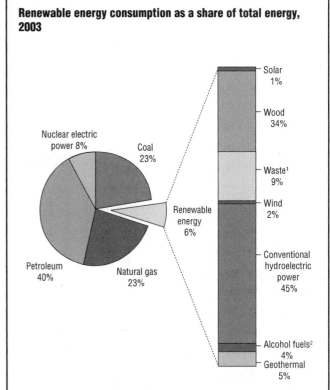

Renewable energy consumption as a share of total energy, 2003

[1]Municipal solid waste, landfill gas, sludge waste, tires, agricultural byproducts, and other biomass.
[2]Ethanol blended into motor gasline.

SOURCE: Adapted from "Figure 10.1. Renewable Energy Consumption by Major Sources: Renewable Energy as Share of Total Energy, 2003," in *Annual Energy Review 2003,* U.S. Department of Energy, Energy Information Administration, Office of Energy Markets and End Use, September 7, 2004, http://www.eia.doe.gov/emeu/aer/pdf/aer.pdf (accessed September 28, 2004)

Wood burning in stoves and fireplaces is another example of direct burning of biomass for energy, in this case heat. In the United States, residential use of wood as fuel generated 359 trillion Btu in 2003. In comparison, the generation of Btu from burning wood in the home in the 1980s ranged from about 850–950 trillion Btu annually according to the Energy Information Administration's *Annual Energy Review 2003*.

Thermochemical Conversion

Thermochemical conversion involves heating biomass in an oxygen-free or low-oxygen atmosphere, transforming the material into simpler substances that can be used as fuels. Products such as charcoal and methanol are produced this way.

Biochemical Conversion

Biochemical conversion uses enzymes, fungi, or other microorganisms to convert high-moisture biomass into either liquid or gaseous fuels. Bacteria convert manure, agricultural wastes, paper, and algae into methane, which is used as fuel. Sewage treatment plants have used anaero-

TABLE 6.1

Energy consumption by source, selected years, 1949–2003

(Quadrillion btu)

Year	Fossil fuels					Nuclear electric power	Hydroelectric pumped storage[5]	Renewable Energy[1]						Electricity net imports	Total[4,6]
	Coal	Coal coke net imports	Natural gas[2]	Petroleum[3,4]	Total			Conventional hydroelectric power	Wood, waste, alcohol[4,6]	Geothermal	Solar	Wind	Total		
1949	11.981	−0.007	5.145	11.883	29.002	0.000	7	1.425	1.549	NA	NA	NA	2.974	0.005	31.982
1950	12.347	0.001	5.968	13.315	31.632	0.000	7	1.415	1.562	NA	NA	NA	2.978	0.006	34.616
1955	11.167	−0.010	8.998	17.255	37.410	0.000	7	1.360	1.424	NA	NA	NA	2.784	0.014	40.208
1960	9.838	−0.006	12.385	19.919	42.137	0.006	7	1.608	1.320	0.001	NA	NA	2.929	0.015	45.087
1965	11.581	−0.018	15.769	23.246	50.577	0.043	7	2.059	1.335	0.004	NA	NA	3.398	(s)	54.017
1970	12.265	−0.058	21.795	29.521	63.522	0.239	7	2.634	1.431	0.011	NA	NA	4.076	0.007	67.844
1971	11.598	−0.033	22.469	30.561	64.596	0.413	7	2.824	1.432	0.012	NA	NA	4.268	0.012	69.289
1972	12.077	−0.026	22.698	32.947	67.696	0.584	7	2.864	1.503	0.031	NA	NA	4.398	0.026	72.704
1973	12.971	−0.007	22.512	34.840	70.316	0.910	7	2.861	1.529	0.043	NA	NA	4.433	0.049	75.708
1974	12.663	0.056	21.732	33.455	67.906	1.272	7	3.177	1.540	0.053	NA	NA	4.769	0.043	73.991
1975	12.663	0.014	19.948	32.731	65.355	1.900	7	3.155	1.499	0.070	NA	NA	4.723	0.021	71.999
1976	13.584	(s)	20.345	35.175	69.104	2.111	7	2.976	1.713	0.078	NA	NA	4.768	0.029	76.012
1977	13.922	0.015	19.931	37.122	70.989	2.702	7	2.333	1.838	0.077	NA	NA	4.249	0.059	78.000
1978	13.766	0.125	20.000	37.965	71.856	3.024	7	2.937	2.038	0.064	NA	NA	5.039	0.067	79.986
1979	15.040	0.063	20.666	37.123	72.892	2.776	7	2.931	2.152	0.084	NA	NA	5.166	0.069	80.903
1980	15.423	−0.035	20.394	34.202	69.984	2.739	7	2.900	2.485	0.110	NA	NA	5.494	0.071	78.289
1981	15.908	−0.016	19.928	31.931	67.750	3.008	7	2.758	2.590	0.123	NA	NA	5.471	0.113	R76.342
1982	15.322	−0.022	18.505	30.232	64.037	3.131	7	3.266	2.615	0.105	NA	NA	5.985	0.100	R73.253
1983	15.894	−0.016	17.357	30.054	63.290	3.203	7	3.527	2.831	0.129	NA	NA	6.488	0.121	R73.101
1984	17.071	−0.011	18.507	31.051	66.617	3.553	7	3.386	2.880	0.165	NA	NA	6.431	0.135	R76.736
1985	17.478	−0.013	17.834	30.922	66.221	4.076	7	2.970	2.864	0.198	NA	NA	6.033	0.140	R76.469
1986	17.260	−0.017	16.708	32.196	66.148	4.380	7	3.071	2.841	0.219	NA	NA	6.132	0.122	R76.782
1987	18.008	0.009	17.744	32.865	68.626	4.754	7	2.635	2.823	0.229	NA	NA	5.687	0.158	R79.225
1988	18.846	0.040	18.552	34.222	71.660	5.587	7	2.334	2.937	0.217	NA	NA	5.489	0.108	R82.844
1989	19.070	0.030	19.712	34.211	73.023	5.602	7	2.837	3.062	0.317	0.055	0.022	6.294	0.037	R84.957
1990	19.173	0.005	19.730	33.553	72.460	6.104	−0.036	3.046	2.662	0.336	0.060	0.029	6.133	0.008	R84.668
1991	18.992	0.010	20.149	32.845	71.996	6.422	−0.047	3.016	2.702	0.346	0.063	0.031	6.158	0.067	R84.595
1992	19.122	0.035	20.835	33.527	73.519	6.479	−0.043	2.617	2.847	0.349	0.064	0.030	5.907	0.087	R85.949
1993	19.835	0.027	21.351	433.841	75.055	6.410	−0.042	2.892	4,R2.803	0.364	0.066	0.031	R6.156	0.095	4,R87.578
1994	19.909	0.058	21.842	34.670	76.480	6.694	−0.035	2.683	2.939	0.338	0.069	0.036	6.065	0.153	89.248
1995	20.089	0.061	22.784	34.553	77.488	7.075	−0.028	3.205	3.068	0.294	0.070	0.033	6.669	0.134	91.221
1996	21.002	0.023	23.197	35.757	79.978	7.087	−0.032	3.590	3.127	0.316	0.071	0.033	7.137	0.137	94.224
1997	21.445	0.046	23.329	36.266	81.086	6.597	−0.041	3.640	3.006	0.325	0.070	0.034	7.075	0.116	94.727
1998	21.656	0.067	22.936	36.934	81.592	7.068	−0.046	3.297	2.835	0.328	0.070	0.031	6.561	0.088	95.146
1999	21.623	0.058	23.010	37.960	82.650	7.610	−0.062	3.268	2.885	0.331	0.069	0.046	6.599	0.099	96.774
2000	22.580	0.065	R23.916	38.404	R84.965	7.862	−0.057	2.811	2.907	0.317	0.066	0.057	6.158	R0.115	R98.905
2001	R21.952	R0.029	R22.906	R38.333	R83.221	R8.033	−0.090	R,P2.201	R2.640	0.311	P0.065	RP0.068	R5.286	0.075	R96.378
2002	R21.980	R0.061	R23.662	R38.401	R84.104	8.143	RP −0.088	RP2.675	R2.791	R0.328	P0.064	RP0.105	RP5.963	0.078	R98.026
2003	P22.707	P0.051	P22.507	P39.074	P84.338	P7.973	P −0.088	P2.779	P2.884	P0.314	P0.063	P0.108	P6.150	P0.022	P98.156

TABLE 6.1

Energy consumption by source, selected years, 1949–2003 [CONTINUED]

(Quadrillion btu)

[1] Electricity net generation from conventional hydroelectric power, geothermal, solar, and wind; consumption of wood, waste, and alcohol fuels; geothermal heat pump and direct use energy; and solar thermal direct use energy.
[2] Natural gas, plus a small amount of supplemental gaseous fuels that cannot be identified separately.
[3] Petroleum products supplied, including natural gas plant liquids and crude oil burned as fuel. Beginning in 1993, also includes ethanol blended into motor gasoline.
[4] Beginning in 1993, ethanol blended into motor gasoline is included in both "Petroleum" and "Wood, waste, alcohol," but is counted only once in total consumption.
[5] Pumped storage facility production minus energy used for pumping.
[6] "Alcohol" is ethanol blended into motor gasoline.
[7] Included in "Conventional hydroelectric power."
R = Revised.
P = Preliminary.
NA = Not available.
(s) = Less than 0.0005 and greater than −0.0005 quadrillion Btu.
Note: Totals may not equal sum of components due to independent rounding.
Web Page: For data not shown for 1951–1969, see http://www.eia.doe.gov/emeu/aer/overview.html.

SOURCE: "Table 1.3. Energy Consumption by Source, Selected Years, 1949–2003 (Quadrillion Btu)," in *Annual Energy Review 2003*, U.S. Department of Energy, Energy Information Administration, Office of Energy Markets and End Use, September 7, 2004, http://www.eia.doe.gov/emeu/aer/pdf/aer.pdf (accessed September 28, 2004)

bic (oxygen-free) digestion for many years to generate methane gas. Small-scale digesters have been used on farms, primarily in Europe and Asia, for hundreds of years. Biogas pits (a biomass-based technology) are a significant source of energy in China.

Another type of biochemical conversion process, fermentation, uses yeast to decompose carbohydrates, yielding ethyl alcohol (ethanol) and carbon dioxide. Sugar crops, grains (corn, in particular), potatoes, and other starchy crops commonly supply the sugar for ethanol production.

Ethanol and Methanol

Ethanol (ethyl alcohol) is a colorless, nearly odorless, flammable liquid derived from fermenting plant material that contains carbohydrates in the form of sugar. Most of the ethanol manufactured for use as fuel in the United States is derived from corn, wood, and sugar. Gasohol is a product formed by mixing ethanol and gasoline. There are three types of gasohol: 10% gasohol, which is a mixture of 10% ethanol and 90% gasoline; 7.7% gasohol, which is at least 7.7% ethanol but less than 10%; and 5.7% gasohol, which is at least 5.7% ethanol but less than 7.7%. The Federal Highway Administration estimated that in 2002 nearly 21 billion gallons of gasohol were used by Americans, up from 17.4 billion gallons in 2001 and 16.3 billion gallons in 2000.

Automobiles can run on gasohol and can be built to run directly on ethanol or on any mixture of ethanol and gasoline. However, ethanol is difficult and expensive to produce in bulk. The development of ethanol as a fuel source may depend more upon the political support of legislators from farming states and a desire for some independence from foreign oil rather than upon savings at the gas pump.

Some scientists believe ethanol made from wood, sawdust, corncobs, or rice hulls could liberate the alcohol fuel industry from its dependence on food crops, such as corn and sugarcane. Worldwide, enough corncobs and rice hulls are left over from annual crop production to produce more than forty billion gallons of ethanol.

Advocates of wood-derived ethanol believe that it could eventually be a sustainable liquid fuel industry that does not rely on pollution-generating fossil fuels. For instance, if new trees were planted to replace those that were cut for fuel, they would be available for later harvesting while at the same time alleviating global warming with their carbon dioxide-processing function. However, other scientists warn that an increased demand for wood for transportation fuels might accelerate the destruction of old-growth forests and endanger ecosystems.

Methanol (methyl alcohol) fuels have also been tested successfully. Using methanol instead of diesel fuel virtually eliminates sulfur emissions and reduces other environmental pollutants usually emitted from trucks and buses. Producing methanol from biofuels, however, is costly.

Burning biofuels in vehicle engines is part of the "carbon cycle" in which the earth's vegetation can, in turn, make use of the products of automobile combustion. (See Figure 6.2.) Automobile combustion generated from fossil fuels, however, contains pollutants. In addition, generating excessive amounts of carbon dioxide from either fossil fuels or biofuels is thought to add to global warming because this gas acts as a "blanket," trapping heat between the earth and the atmosphere.

Municipal Waste Recovery

Each year millions of tons of garbage are buried in landfills and city dumps. This method of disposal is becoming increasingly costly as many landfills across the nation near capacity. Many communities discovered that they could solve both problems—cost and capacity—by constructing waste-to-energy plants. Not only is the garbage burned and reduced in volume by 90%, but also energy in the form of steam or electricity is generated in a cost-effective way, and the potential energy benefit is significant. Use of municipal waste as fuel has increased steadily since the 1980s. According to EIA figures, municipal waste (including landfill gas, sludge waste, tires, and agricultural by-products) generated 88 trillion Btu of energy in 1981, which grew to 558 trillion Btu by 2003.

The two most common waste-to-energy plant designs are the mass burn (also called direct combustion) system and the refuse derived fuel (RDF) system.

MASS BURN SYSTEMS. Most waste-to-energy plants in the United States use the mass burn system. This system's advantage is that the waste does not have to be sorted or prepared before burning, except for removing obviously noncombustible, oversized objects. The mass burn eliminates expensive sorting, shredding, and transportation machinery that may be prone to break down.

In mass burn systems, waste is carried to the plant in trash trucks and dropped into a storage pit. Large overhead cranes lift the garbage into a furnace feed hopper that controls the amount and rate of waste that is fed into the furnace. Next, the garbage is moved through a combustion zone so that it burns to the greatest extent possible. The burning waste produces heat, and that heat is used to produce steam. The steam can be used directly for industrial needs or can be sent through a turbine to power a generator to produce electricity.

REFUSE DERIVED FUEL (RDF) SYSTEMS. RDF systems process waste to remove noncombustible objects and to create homogeneous and uniformly sized fuel. Large items such as bedsprings, dangerous materials, and flammable liquids are removed by hand. The trash is then shredded and carried to a screen to remove glass, rocks,

FIGURE 6.2

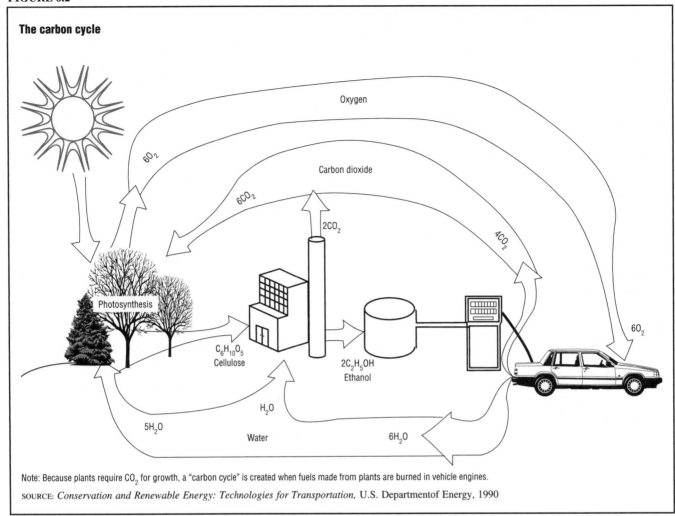

The carbon cycle

Photosynthesis

$6O_2$

Oxygen

$6CO_2$

Carbon dioxide

$2CO_2$

$4CO_2$

$6O_2$

$C_6H_{10}O_5$
Cellulose

$2C_2H_5OH$
Ethanol

H_2O

$5H_2O$

Water

$6H_2O$

Note: Because plants require CO_2 for growth, a "carbon cycle" is created when fuels made from plants are burned in vehicle engines.

SOURCE: *Conservation and Renewable Energy: Technologies for Transportation,* U.S. Departmentof Energy, 1990

and other material that cannot be burned. The remaining material is usually sifted a second time with an air separator to yield fluff. The fluff is sent to storage bins before being burned, or it can be compressed into pellets or briquettes for long-term storage. This fuel can be used as an energy source by itself in a variety of systems, or it can be used with other fuels, such as coal or wood.

PERFORMANCE OF WASTE-TO-ENERGY SYSTEMS. Most waste-to-energy systems can produce two to four pounds of steam for every pound of garbage burned. A 1,000-ton-per-day mass burn system will burn an average of 310,250 tons of trash each year and will recover two trillion Btu of energy. In addition, the plant will emit 96,000 tons of ash (32% of waste input) for landfill disposal. An RDF plant produces less ash but sends almost the same amount of waste to the landfill because of the noncombustibles that accumulate in the separation process before burning.

DISADVANTAGES OF WASTE-TO-ENERGY PLANTS. The major problem with increasing the use of municipal waste-to-energy plants is their effect on the environment. The emission of particles into the air is partially controlled by

electrostatic precipitators, and many gases can be eliminated by proper combustion techniques. There is concern, however, about the amounts of dioxin (a very dangerous air pollutant) and other toxins that are often emitted from these plants. Noise from trucks, fans, and processing equipment at these plants can also be unpleasant for nearby residents.

Landfill Gas Recovery

Landfills contain a large amount of biodegradable matter that is compacted and covered with soil. Bacteria called methanogens thrive in this oxygen-depleted environment. They metabolize the biodegradable matter in the landfill, producing methane gas and carbon dioxide as by-products. In the past, as landfills aged, these gases built up and leaked out. This gas leakage prompted some communities to drill holes in landfills and burn off the methane to prevent dangerously large amounts from exploding.

The energy crisis of the 1970s made landfill methane gas an energy resource too valuable to waste, and efforts were made to find an inexpensive way to tap the gas. The first landfill gas recovery site was finished in 1975 at the Palos Verdes Landfill in Rolling Hills Estates, California.

In a typical operation, garbage is allowed to decompose for several months. When a sufficient amount of methane gas has developed, it is piped out to a generating plant, where it is turned into electricity. In its purest form, methane gas is equivalent to natural gas and can be used in exactly the same way. Depending on the extraction rates, most sites can produce gas for about 20 years. The advantages of tapping gas from a landfill go beyond the energy provided by the methane, as extraction reduces landfill odors and the chances of explosions.

HYDROPOWER

Hydropower, the energy that comes from the natural flow of water, is the world's largest renewable energy source. The energy of falling water or flowing water is converted into mechanical energy and then to electrical energy. In the past, flowing water turned waterwheels to grind grain or turn saws, but today flowing water is used to turn modern turbines. Hydropower is a renewable, nonpolluting, and reliable energy source.

Advantages and Disadvantages of Using Hydropower Energy

At present, hydropower is the only means of storing large quantities of electrical energy for almost instant use. This is done by holding water in a large reservoir behind a dam, with a hydroelectric power plant below. The dam creates a height from which water flows. The fast-moving water pushes the turbine blades that turn the rotor part of the electric generator. Whenever power is needed at peak times, the valves are opened, and turbine generators quickly produce power.

Nearly all the best sites for large hydropower plants are already being put to use in the United States. Small hydropower plants are expensive to build but eventually become cost-efficient because of their low operating costs. One of the disadvantages of small hydropower generators is their reliance on rain and melting snow to fill reservoirs because some years bring drought conditions. Additionally, U.S. environmental groups strongly protest the construction of new dams in America. Ecologists express concern that dams ruin streams, dry up waterfalls, and interfere with aquatic life habitats.

New Directions in Hydropower Energy

The United States and Europe have developed a major proportion of their hydroelectric potential. Large-scale hydropower development has slowed considerably in the United States. The last federally funded hydropower dam constructed in the United States was the Corps of Engineers' Richard B. Russell Dam and Lake, which is located on the Savannah River and borders South Carolina and Georgia. The project was authorized in 1966 and completed in 1986. However, expansion and efficiency improvements at existing dams still offer significant potential for additional hydropower capacity and energy. Until recently in the United States, dams were usually funded entirely with federal monies. Since 1986, however, local governments must contribute half of the cost of any new dam proposed in the United States. Hydropower's contribution to U.S. energy generation should remain relatively constant, although existing sites can become more efficient as new generators are added. Any new major supplies of hydroelectric power for the United States will likely come from Canada.

Most of the new development in hydropower is occurring in the Third World, as developing nations see it as an effective method of supplying power to growing populations. Most of these hydropower-development programs are massive public-works projects requiring huge amounts of money, which is mostly borrowed from the developed world. Third World leaders believe that hydroelectric dams are worth the cost and potential environmental threats because they bring cheap electric power to their citizenry.

GEOTHERMAL ENERGY

Since ancient times, humans have exploited the earth's natural hot water sources. Although bubbling hot springs became public baths as early as ancient Rome, using hot water and underground steam to produce power is a relatively recent development. Electricity was first generated from natural steam in Italy in 1904. The world's first steam power plant was built in 1958 in a volcanic region of New Zealand. A field of twenty-eight geothermal power plants covering thirty square miles in northern California was completed in 1960.

What Is Geothermal Energy?

Geothermal energy is the natural, internal heat of the earth trapped in rock formations deep underground. Only a fraction of this vast storehouse of energy can be extracted, usually through large fractures in the earth's crust. Hot springs, geysers, and fumaroles (holes in or near volcanoes from which vapor escapes) are the most easily exploitable sources of geothermal energy. (See Figure 6.3.) Geothermal reservoirs provide hot water or steam that can be used for heating buildings, processing food, and generating electricity.

To produce power from a geothermal energy source, pressurized steam or hot water is extracted from the earth and directed toward turbines. The electricity produced by the turbines is then fed into a utility grid and distributed to residential and commercial customers.

Types of Geothermal Energy

Like most natural energy sources, geothermal energy is usable only when it is concentrated in one spot, in this

FIGURE 6.3

Cross-section of the Earth showing source of geothermal energy

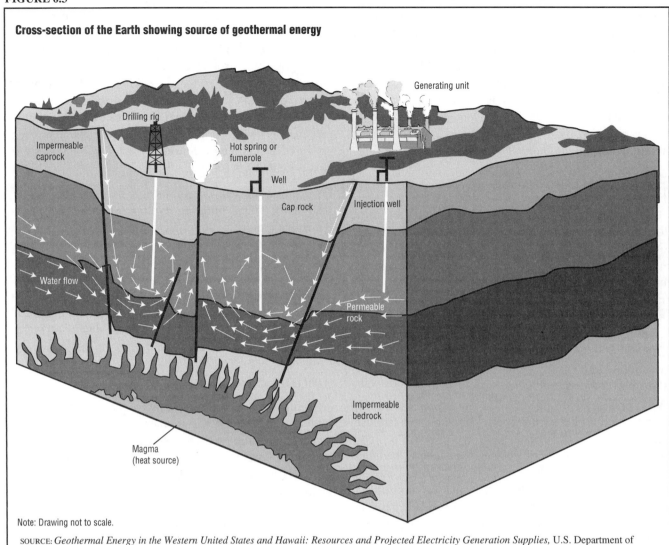

Note: Drawing not to scale.

SOURCE: *Geothermal Energy in the Western United States and Hawaii: Resources and Projected Electricity Generation Supplies,* U.S. Department of Energy, Energy Information Administration, 1991

case in what is called a "thermal reservoir." The four basic categories of thermal reservoirs are hydrothermal reservoirs, dry rock, geopressurized reservoirs, and magma resources. Most of the known reservoirs for geothermal power in the United States are located west of the Mississippi River, and the highest-temperature geothermal resources occur mostly west of the Rocky Mountains. According to the Energy Information Administration, in 2002 geothermal resources produced nearly 13.4 billion kilowatt hours, which is a little less than 4% of the energy generated by renewable sources.

HYDROTHERMAL RESERVOIRS. Hydrothermal reservoirs consist of a heat source covered by a permeable formation through which water circulates. Steam is produced when hot water boils underground and some of the steam escapes to the surface under pressure. Once at the surface, impurities and tiny rock particles are removed, and the steam is piped directly to the electrical generating station. These systems are the cheapest and simplest form of geo-

thermal energy. The Geysers, ninety miles north of San Francisco, California, are the most famous example of this type. The Geysers Geothermal Field is the world's largest source of geothermal power, according to the Energy Information Administration.

DRY ROCK. Dry rock formations are the most common geothermal source, especially in the West. To tap this source of energy, water is injected into hot rock formations that have been fractured and the resulting steam or water is collected.

GEOPRESSURIZED RESERVOIRS. Geopressurized reservoirs are sedimentary formations containing hot water and methane gas. Supplies of geopressurized energy remain uncertain, and drilling is expensive. Scientists hope that advancing technology will eventually permit the commercial exploitation of the methane content in these reservoirs.

MAGMA RESOURCES. Magma resources are found from ten thousand to thirty-three thousand feet below the earth's surface, where molten or partially liquefied rock is located. Because magma is so hot, ranging from 1,650 to 2,200 degrees Fahrenheit, it is a good geothermal resource. The process for extracting energy from magma is still in the experimental stages.

Domestic Production of Geothermal Energy

Geothermal energy ranked third in renewable energy production in the United States in 2003, after biomass (wood, waste, alcohol) and hydroelectric power. (See Table 6.1.) According to the International Geothermal Association, in 2002 the United States had 28% of the installed geothermal generating capacity of the world, but most of the easily exploited geothermal reserves in the United States have already been developed. In addition, utility companies and independent power producers are arguing over who should build additional generating capacity and what prices should be paid for the power. Continued growth in the American market depends on the regulatory environment, oil price trends, and the success of unproven technologies for economically exploiting some of the presently inaccessible geothermal reserves.

International Production of Geothermal Energy

Since 1970, worldwide geothermal electrical generating capacity has more than tripled. According to the EIA's *Annual Energy Review 2003*, geothermal energy made up about 1.2% of world electrical production in 2002. Geothermal sources worldwide produce little more than the energy output of ten average-size coal-fired power plants.

World geothermal reserves are immense but unevenly distributed. They fall mostly in seismically active areas at the margins or borders of the earth's nine tectonic plates. Currently, exploited reserves represent only a small fraction of the overall potential—many countries are believed to have in excess of 100,000 megawatts of geothermal energy available.

The World Geothermal Congress (WGC), with representation by delegates from sixty countries, met in Kyushu and Tohoku, Japan in 2000. At that meeting the WGC noted that nearly 90% of homes and other buildings in Iceland are heated by geothermal waters, and approximately 26% of electrical power generation in the Philippines comes from geothermal steam. The WGC supports the use of geothermal energy; one of its goals is to replicate such high use of geothermal resources in other countries. The next meeting of WGC is in 2005.

Disadvantages of Geothermal Energy

Geothermal plants must be built near a geothermal source, are not very efficient, produce unpleasant odors from sulfur released in processing, generate noise, are inaccessible for most states, release potentially harmful pollutants (hydrogen sulfide, ammonia, and radon), and

release poisonous arsenic or boron often found in geothermal waters. Serious environmental concerns have been raised over the release of chemical compounds, the potential contamination of water sources, the collapse of the land surface around the area from which the water is being drained, and potential water shortages resulting from massive withdrawals of water.

WIND ENERGY

Winds are created by the uneven heating of the atmosphere by the sun, the irregularities of the earth's surface, and the rotation of the earth. Winds are strongly influenced by local terrain, water bodies, weather patterns, vegetation, and other factors. Wind flow, when "harvested" by wind turbines, can be used to generate electricity.

Early windmills produced mechanical energy to pump water and run grain and sawmills. In the late 1890s, Americans began experimenting with wind power to generate electricity. Their early efforts produced enough electricity to light one or two modern light bulbs.

Compared to the pinwheel-shaped farm windmills that can still be seen dotting the American rural landscape, today's state-of-the-art wind turbines look more like airplane propellers. Their sleek, high-tech fiberglass design and aerodynamics allow them to generate an abundance of electricity while they also produce mechanical energy and heat.

Beginning in the 1990s, industrial and developing countries alike have started using wind power as a source of electricity to complement existing power sources and to bring electricity to remote regions. Wind turbines cost less to install per unit of kilowatt capacity than either coal or nuclear facilities. After installing a windmill, there are few additional costs, as the fuel (wind) is free.

Wind speeds are generally highest and most consistent in mountain passes and along coastlines. Europe has the greatest coastal wind resources, and clusters of wind turbines, or wind farms, are being developed there and in Asia. Denmark, the Netherlands, China, and India are especially interested in fostering the development of domestic wind industries (*International Energy Outlook 2004*, Energy Information Administration). As of 2004, electricity-producing wind turbines operate in ninety-five countries. In the United States it is estimated that sufficient wind energy is available to provide more than one trillion kilowatt-hours of electricity annually, or about 27% of the total used in 2003.

Domestic Energy Production by Wind Turbines

According to the EIA, wind is the world's fastest-growing renewable energy source (*International Energy Outlook 2004*). Although wind power has not been adopted

FIGURE 6.4

Wind capacity map, 2003

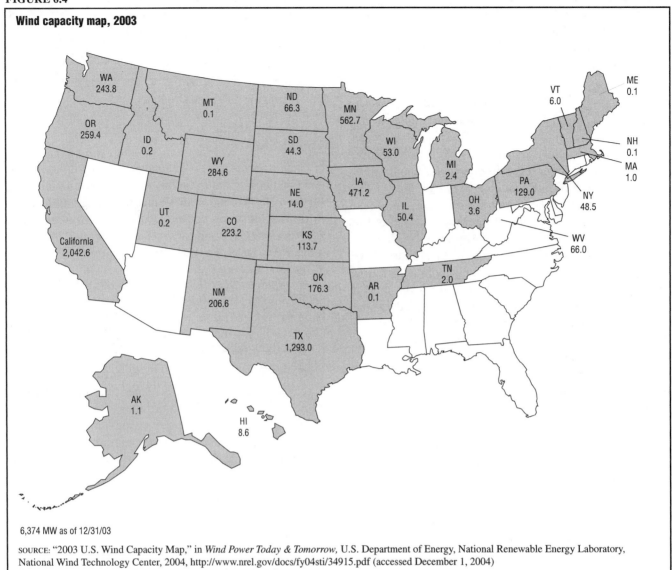

6,374 MW as of 12/31/03

SOURCE: "2003 U.S. Wind Capacity Map," in *Wind Power Today & Tomorrow,* U.S. Department of Energy, National Renewable Energy Laboratory, National Wind Technology Center, 2004, http://www.nrel.gov/docs/fy04sti/34915.pdf (accessed December 1, 2004)

widely in the United States, U.S. companies export turbines to Spain, the Netherlands, Great Britain, India, and China.

In the United States, the wind industry began in California in 1981 with the erection of 144 relatively small turbines capable of generating a combined total of seven megawatts of electricity. Within a year the number of turbines had increased ten times, and by 1986 they had multiplied one-hundred-fold. In *Annual Energy Outlook 2004,* the EIA projects wind power capacity in the United States to grow more than three-fold from 2003 to 2025.

Wind technology exploded in California in the 1980s, where about 95% of the installed wind capacity in the United States used to be located. During 1998 and 1999, however, wind farm activity expanded into other states; less than 32% of new wind power construction was located in California. This increasing activity outside California was motivated by financial incentives (such as the wind-energy-production tax credit), regulatory incentives, and

state mandates (in Iowa and Minnesota). In 1999 Iowa, Minnesota, and Texas each had capacity additions exceeding one hundred megawatts. According to the American Wind Energy Association, by 2004 these three states had been joined by nine others in exceeding one hundred megawatts of installed capacity. These twelve states—Washington, Oregon, California, Wyoming, Colorado, New Mexico, Texas, Iowa, Minnesota, Oklahoma, Kansas, and Pennsylvania—contain 94% of the U.S. wind energy potential. The wind power generation capacity of the United States in 2003 is shown in Figure 6.4. The total installed generating capacity of the U.S. is 6,374 megawatts, and wind power plants operate in thirty-two states.

Refinements in wind-turbine technology may enable a substantial portion of the nation's electricity to be produced by wind energy. Use of this technology is being encouraged by the initiative "Wind Powering America," which was announced in June 1999 by the U.S. Secretary of Energy. The stated goals of the program are to have eighty thousand megawatts of wind power generation capacity in place by 2020 and to have wind power provide 5% of the nation's electricity generation.

An added incentive to developing wind technology is continuing tax credits. The wind-energy-production tax credit provided by the Energy Policy Act (EPACT) of 1992 was scheduled to expire in 1999 but was extended to the end of 2003. A provision in the 2003 EPACT bill includes a bipartisan plan for extending the tax credit through 2006.

International Development of Wind Energy

During the decade following the 1973 oil embargo, more than ten thousand wind machines were installed worldwide, ranging in size from portable units to multi-megawatt turbines. In the villages of developing nations, small wind turbines recharge batteries and provide essential services. In China small wind turbines allow people to watch their favorite television shows, an activity that has increased wind energy demand. In fact, in 2001 China was the world's largest manufacturer of small wind turbines.

Global wind-power-generating capacity was about 39,000 megawatts in 2003, up from 23,300 megawatts in 2001, 7,200 megawatts in 1997, and 3,000 megawatts in 1993. Germany, Spain, and Denmark are the fastest growing wind producers in the world, and the United Kingdom, Ireland, and Portugal all are experiencing a surge in installed wind capacity ("The Current Status of the Wind Industry," European Wind Energy Association, http://www.ewea.org/ [accessed January 14, 2005]).

Interest in wind energy has been driven, in part, by the declining cost of capturing wind energy. From more than thirty-eight cents per kilowatt-hour in 1980, wind energy prices declined to about four cents per kilowatt-hour in 2002 for new turbines at sites with strong winds (Lester R. Brown, "Wind Power Set to Become World's Leading Energy Source," Earth Policy Institute, June 25, 2003). Decreasing costs could make wind power competitive with gas and coal power plants, even before considering wind's environmental advantages.

Advantages and Disadvantages of Wind Energy

The main problem with wind energy is that the wind does not always blow. Some people object to the whirring noise of wind turbines or do not like to see wind turbines clustered in mountain passes and along shorelines because they interfere with scenic views. Environmentalists have charged that wind turbines are responsible for the loss of thousands of endangered birds that fly into the blades, as birds frequently use windy passages in their travel patterns.

However, generating electricity with wind offers many environmental advantages. Wind farms do not emit climate-altering carbon dioxide, acid-rain-forming pollu-tants, respiratory irritants, or nuclear waste. Because wind farms do not require water to operate, they are especially well-suited to semi-arid and arid regions.

SOLAR ENERGY

Ancient Greek and Chinese civilizations used glass and mirrors to direct the sun's rays to start fires. Solar energy (energy from the sun) is a renewable, widely available energy source that does not generate greenhouse gases or radioactive waste. Solar-powered cars have competed in long-distance races, and solar energy has been used routinely for many years to power spacecraft. Although many people consider solar energy a product of the space age, architectural researchers at the Massachusetts Institute of Technology built the first solar house in 1939.

Solar radiation is nearly constant outside the earth's atmosphere, but the amount of solar energy reaching any point on earth varies with changing atmospheric conditions, such as clouds and dust, and the changing position of the earth relative to the sun. In the United States, exposure to the sun's rays is greatest in the West and Southwest regions. Nevertheless, almost all U.S. regions have solar resources that can be used. (See Figure 6.5.)

Passive and Active Solar Energy Collection Systems

Passive solar energy systems, such as greenhouses or windows with a southern exposure, use heat flow, evaporation, or other natural processes to collect and transfer heat. (See Figure 6.6.) They are considered the least costly and least difficult solar systems to implement.

Active solar systems use mechanical methods to control the energy process. (See Figure 6.6.) They require collectors and storage devices as well as motors, pumps, and valves to operate the systems that transfer heat. Collectors consist of an absorbing plate that transfers the sun's heat to a working fluid (liquid or gas), a translucent cover plate that prevents the heat from radiating back into the atmosphere, and insulation on the back of the collector panel to further reduce heat loss. Excess solar energy is transferred to a storage facility so it may be used to provide power on cloudy days. In both active and passive systems, the conversion of solar energy into a form of power is made at the site where it is used. The most common and least expensive active solar systems are used for heating water.

Solar Thermal Energy Systems

A solar thermal energy system uses intensified sunlight to heat water or other fluids to temperatures of more than 750 degrees Fahrenheit. Mirrors or lenses constantly track the sun's position and focus its rays onto solar receivers that contain fluid. Solar heat is transferred to the water, which in turn powers an electric generator. In a distributed solar thermal system, the collected energy powers

FIGURE 6.5

Solar resources

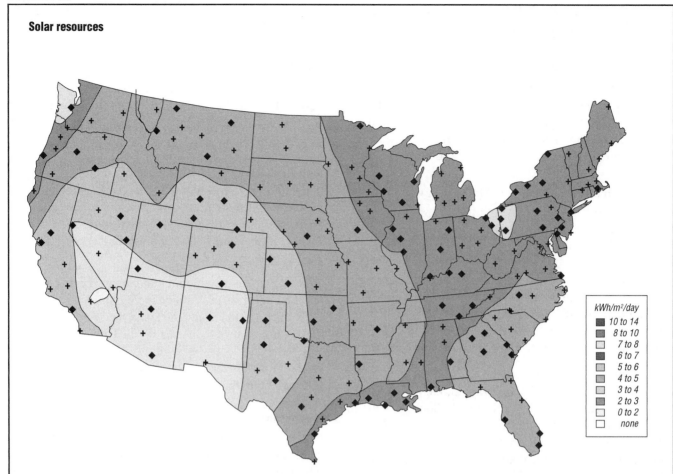

SOURCE: Larry Prete, "Figure 2. Annual Average Daily Total Solar Resources," in *Transmission Pricing Issues for Electricity Generation from Renewable Resources: Background,* U.S. Department of Energy, Energy Information Administration, March 1999, http://www.eia.doe.gov/cneaf/solar.renewables/ rea_issues/html/background.html (accessed November 18, 2004)

irrigation pumps, providing electricity for small communities or capturing normally wasted heat from the sun in industrial areas. In a central solar thermal system, the energy is collected at a central location and used by utility networks for a large number of customers.

Other solar thermal energy systems include solar ponds and trough systems. Solar ponds are lined ponds filled with water and salt. Because salt water is denser than fresh water, the salt water on the bottom absorbs the heat, and the fresh water on top keeps the salt water contained and traps the heat. Trough systems use U-shaped mirrors to concentrate the sunshine on water or oil-filled tubes.

Photovoltaic Conversion Systems

The photovoltaic (PV) cell solar energy system converts sunlight directly into electricity without the use of mechanical generators. PV cells have no moving parts, are easy to install, require little maintenance, do not pollute the air, and can last up to twenty years. PV cells are commonly used to power small devices, such as watches or calculators. They are also being used on a larger scale

to provide electricity for rural households, recreational vehicles, and businesses. Solar panels using photovoltaic cells have generated electricity for space stations and satellites for many years.

Since PV systems produce electricity only when the sun is shining, a backup energy supply is required. PV cells produce the most power around noon, when sunlight is the most intense. A photovoltaic system typically includes storage batteries that provide electricity during cloudy days and at night.

The use of photovoltaic technology is expanding both in the United States and abroad. PV systems have low operating costs because there are no turbines or other moving parts, and maintenance is minimal. PV cell systems are nonpolluting and silent and can be operated by computer. Above all, the fuel source (sunshine) is free and plentiful. The main disadvantage of photovoltaic cell energy systems is the initial cost. Although the price has fallen considerably, PV cells are still too expensive for widespread use. PV systems also use some toxic materials, which may cause environmental problems.

FIGURE 6.6

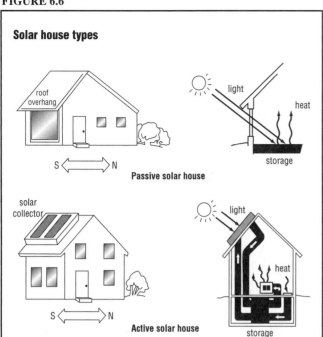

Solar house types

Passive solar house

Active solar house

SOURCE: U.S. Department of Energy

FIGURE 6.7

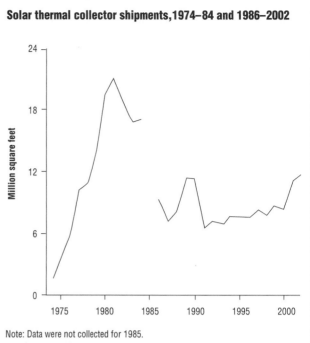

Solar thermal collector shipments,1974–84 and 1986–2002

Note: Data were not collected for 1985.

SOURCE: Adapted from "Figure 10.3. Solar Thermal Collector Shipments by Type, Price, and Trade: Total Shipments, 1974–1984 and 1986–2002," in *Annual Energy Review 2003,* U.S. Department of Energy, Energy Information Administration, Office of Energy Markets and End Use, September 7, 2004, http://www.eia.doe.gov/emeu/aer/pdf/aer.pdf (accessed September 28, 2004)

Solar Energy Usage

Because it is difficult to measure directly the use of solar energy, shipments of solar equipment can be used as an indicator of use. From a high of eighty-four low-temperature solar collector manufacturers in 1979, this number dropped to only thirteen manufacturers in 1999. Total shipments of solar thermal collectors peaked in 1981 at more than 21 million square feet, fell to a low of 6.6 million square feet in 1991, and rose again to 11.7 million square feet in 2002. (See Figure 6.7.)

Based on 2002 figures most of the solar thermal collectors sold are for residential purposes (see Figure 6.8) and most are sold in sunbelt states. The majority of solar thermal collectors shipped in 2002 were used for heating swimming pools, and a smaller percentage for hot water. (See Figure 6.9.) The market for solar energy space heating has virtually disappeared. Only a small proportion of solar thermal collectors are used for commercial purposes, though some state and municipal power companies have added solar energy systems as adjuncts to their regular power sources during peak hours.

Solar Power as an International Rural Solution

Rural areas are more expensive to serve with energy than cities, and electrification has been slow to reach many people in rural areas of developing countries. In the United States, it was only in 1935, after the Rural Electrification Administration provided low-cost financing to rural electric cooperatives, that most farmers received power. In places such as western China, the Himalayan foothills, and the Amazon basin, the cost of connecting new rural cus-

tomers to electricity grids remains very high. Furthermore, state-owned power systems have been poorly managed in many countries. This has left many national power systems all but bankrupt, and blackouts have become common.

In India blackouts are so common that many factories and other businesses have, at great expense, set up their own private systems, using natural gas, propane, or fuel oil. Although rural families do not have access to those systems, they do have sunlight. In most tropical countries, considerable sunlight falls on rooftops. Electricity produced by solar photovoltaic cells was initially too expensive—as much as a thousand times more than that from conventional plants—but it has continually fallen in price.

Future Development Trends

Interest in photovoltaic solar energy systems is particularly high in rural and remote areas where it is impractical to extend traditional electrical power lines. In some remote areas, PV cells are used as independent power sources for communications or for the operation of water pumps or refrigerators.

Although solar power still costs more than three times as much as fossil fuel energy, utilities could turn to solar energy to provide "peaking power" on extremely hot or cold days. In the long run, some people believe that build-

FIGURE 6.8

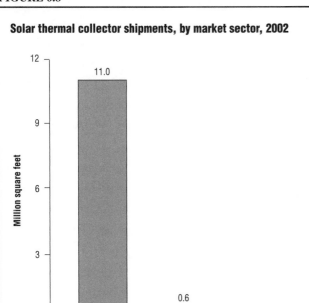

Solar thermal collector shipments, by market sector, 2002

SOURCE: Adapted from "Figure 10.4. Solar Thermal Collector Shipments by End Use, Market Sector, and Type, 2002: Market Sector," in *Annual Energy Review 2003*, U.S. Department of Energy, Energy Information Administration, Office of Energy Markets and End Use, September 7, 2004, http://www.eia.doe.gov/emeu/aer/pdf/aer.pdf (accessed September 28, 2004)

FIGURE 6.9

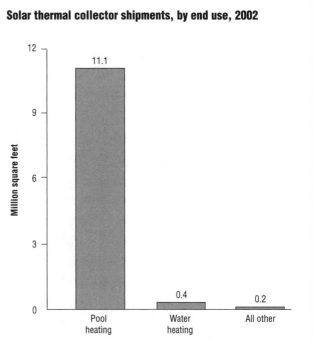

Solar thermal collector shipments, by end use, 2002

SOURCE: Adapted from "Figure 10.4. Solar Thermal Collector Shipments by End Use, Market Sector, and Type, 2002: End Use," in *Annual Energy Review 2003*, U.S. Department of Energy, Energy Information Administration, Office of Energy Markets and End Use, September 7, 2004, http://www.eia.doe.gov/emeu/aer/pdf/aer.pdf (accessed September 28, 2004)

ing solar energy systems to provide peak power capacity would be cheaper than building the new and expensive diesel fuel generators that are now used.

Advantages and Disadvantages of Solar Energy

The primary advantage of solar energy is its inexhaustible supply, while its primary disadvantage is its reliance on a consistently sunny climate to provide continuous electrical power, which is only possible in limited areas. In addition, a large amount of land area is necessary for the most efficient collection of solar energy by electricity plants. Experts estimate that a new thermal energy plant would have a higher cost of production than a conventional coal-fired plant.

POWER FROM THE OCEAN

The potential power of the world's oceans is unknown. Because the ocean is not as easily controlled as a river or water that is directed through canals into turbines, unlocking that potential power is far more challenging. Three ideas being considered are tidal plants, wave power, and ocean thermal energy conversion (OTEC).

Tidal Power

The tidal plant uses the power generated by the tidal flow of water as it ebbs, or flows back out to sea. A minimum tidal range of three to five yards is generally consid-

ered necessary for an economically feasible plant. (The tidal range is the difference in height between consecutive high and low tides.) Canada has built a small 40-megawatt unit at the Bay of Fundy, with its fifteen-yard tidal range, the largest range in the world, and is considering building a larger unit there. The largest existing tidal facility is the 240-megawatt plant at the La Rance estuary in northern France, built in 1965. Russia has a small 400-kilowatt plant near Murmansk, close to the Barents Sea. The world's first offshore tidal energy turbine near Devon, England, began producing energy in 2003.

Wave Energy

Norway has two operating wave power stations at Toftestallen on its Atlantic coast. These systems were the first significant oscillating water column (OWC) systems and they work like this: The arrival of a wave forces water up a hollow sixty-five-foot tower, displacing the air already in the tower. This air rushes out of the top through a turbine. The rotors of that turbine then spin, generating electricity. When the wave falls back and the water level falls, air is sucked back in through the turbine, again generating electricity. OWCs have been built and tested in Japan, Norway, India, China, Scotland, and Portugal.

A second type of wave energy power plant uses the overflow of high waves. As the wave splashes against the top of a dam, some of the water goes over and is trapped in a reservoir on the other side. The water is then directed through a turbine as it flows back to the sea.

These two kinds of plants are experimental. Several projects are under way in Japan and the Pacific region to determine a way to use the potential of the huge waves of the Pacific. Although considerable progress has been made in the research and development of this technology, several challenging engineering problems remain to be solved.

Ocean Thermal Energy Conversion (OTEC)

Ocean thermal energy conversion, or OTEC, uses the temperature difference between the ocean's warm surface water and the cooler water in its depths to produce heat energy that can power a heat engine to produce electricity. OTEC systems can be installed on ships, barges, or off-shore platforms with underwater cables that transmit electricity to shore.

HYDROGEN: A FUEL OF THE FUTURE?

Hydrogen, the lightest and most abundant chemical element, is the ideal fuel from the environmental point of view. Its combustion produces only water vapor, and it is entirely carbon-free. Three-quarters of the mass of the universe is hydrogen, so in theory the supply is ample. However, the combustible form of hydrogen is a gas and is not found in nature. The many compounds containing hydrogen—water, for example—cannot be converted into pure hydrogen without the expenditure of energy. The amount of energy that would be required to make gas is about the same as the amount of energy that would be obtained by the combustion of the hydrogen. Therefore, with today's technology, little or nothing could be gained from an energy point of view.

Scientists, however, are researching ways to produce hydrogen gas economically. Whether this will come from fusion, solar energy, or elsewhere is not possible to predict now. Scientists have considered the possibility of a transition to hydrogen for more than a century, and today many see hydrogen as the logical "third-wave" fuel, with hydrogen gas following oil, just as oil replaced coal decades ear-lier. For now, however, widespread use of hydrogen as fuel is purely theoretical.

Research into the use of hydrogen as a fuel got a boost when President George W. Bush announced a hydrogen fuel initiative in his 2003 State of the Union address. By the end of 2004, the Department of Energy had awarded $75 million in research grants in support of this initiative.

FUTURE TRENDS IN U.S. RENEWABLE ENERGY USE

In *Annual Energy Outlook 2004*, the EIA forecasted that total renewable fuel consumption, including ethanol for transportation, will increase by 1.9% per year from 2002 to 2025. About 60% of the projected demand for renewable fuel in 2025 will be for electricity generation. In 2002 renewable energy provided 5.8 quadrillion Btu, which is projected to increase to 9 quadrillion Btu in 2025. Renewable fuel is expected to remain a small contributor to overall electricity generation, rising only from 9% of the total generation in 2002 to 9.1% in 2025.

Hydropower production is projected to rise only slightly from 260 billion kilowatt-hours in 2002 to 309 billion kilowatt-hours in 2025. The production of other renewables should increase steadily. (See Figure 6.10.) For example, significant increases are projected for both geothermal energy and wind power capacity from 2002 to 2025. The EIA projected that wind power will increase from 4.8 gigawatts in 2002, to 8 gigawatts in 2010, and 16 gigawatts in 2025. The EIA projected that high-output geothermal capacity will increase from 13 billion kilowatt-hours in 2002 to 47 billion in 2025. This, however, will provide less than 1% of the nation's electricity needs.

Municipal solid waste and landfill gas energy production is expected to increase by nine billion kilowatt-hours from 2002 to 2025. The largest source of renewable generation (not including hydropower) is biomass, which is projected to more than double from 2002 to 2025. (See Figure 6.11.) Solar energy is not expected to contribute much to centrally generated electricity.

FIGURE 6.10

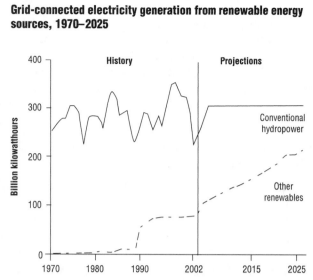

Grid-connected electricity generation from renewable energy sources, 1970–2025

SOURCE: "Figure 77. Grid-Connected Electricity Generation from Renewable Energy Sources, 1970–2025 (Billion Kilowatthours)," in *Annual Energy Outlook 2004,* U.S. Department of Energy, Energy Information Administration, Office of Integrated Analysis and Forecasting, January 2004, http://tonto.eia.doe.gov/FTPROOT/ forecasting/0383(2004).pdf (accessed November 16, 2004)

FIGURE 6.11

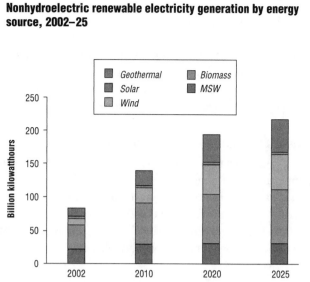

Nonhydroelectric renewable electricity generation by energy source, 2002–25

SOURCE: "Figure 78. Nonhydroelectric Renewable Electricity Generation by Energy Source, 2002–2025 (Billion Kilowatthours)," in *Annual Energy Outlook 2004,* U.S. Department of Energy, Energy Information Administration, Office of Integrated Analysis and Forecasting, January 2004, http://tonto.eia.doe.gov/FTPROOT/ forecasting/0383(2004).pdf (accessed November 16, 2004)

CHAPTER 7
ENERGY RESERVES—OIL, GAS, COAL, AND URANIUM

Fossil fuels and uranium are nonrenewable resources. Nonrenewable resources are defined as concentrations of solid, liquid, or gaseous hydrocarbons that occur naturally in or near the earth's surface. These resources must be currently or potentially recoverable for economic use. They are formed much more slowly than they are consumed, so they are considered nonrenewable. Knowing estimates of the recoverable quantities of crude oil, natural gas, coal, and uranium resources in the United States and worldwide is essential to the development, implementation, and evaluation of national energy policies and legislation since these resources are finite (can be used up). In the United States, Congress requires the Department of Energy (DOE) to prepare estimates of energy reserves.

Proved reserves are reserves from known locations that geological and engineering data demonstrate, with reasonable certainty, to be recoverable with current technological means and economic conditions. Undiscovered recoverable resources are quantities of fuel that are thought to exist in favorable geologic settings. These resources would be feasible to retrieve with existing technological means, although they may not be feasible to recover under current economic conditions.

CRUDE OIL

From 1992 through 2002, total U.S. crude oil proved reserves declined through 1996, rose then fell in 1997 and 1998, and have been climbing since the dip in 1998. On December 31, 2002, crude oil reserves were at 22.7 billion barrels—near the 1997 reserves level. Together, Texas, Alaska, California, and the Gulf of Mexico offshore areas accounted for 79% of U.S. proved reserves in 2002. (See Table 7.1.) Of these four regions, Texas, the Gulf of Mexico offshore areas, and California had increases in crude oil proved reserves in 2002. Alaska reported a decline.

Proved reserves of crude oil and natural gas rose in 1970 with the inclusion of Alaska's North Slope oil fields.

TABLE 7.1

Proved reserves of crude oil, by selected states and state subdivisions, December 31, 2002

Area	Percent of U.S. oil reserves
Texas	22
Alaska	21
Gulf of Mexico federal offshore	20
California	16
Area total	**79**

SOURCE: *U.S. Crude Oil, Natural Gas, and Natural Gas Liquids Reserves 2002 Annual Report,* U.S. Department of Energy, Energy Information Administration, Office of Oil and Gas, December 2003, http://www.eia.doe .gov/pub/oil_gas/natural_gas/data_publications/crude_oil_natural_gas _reserves/current/pdf/arr.pdf (accessed November 17, 2004)

Since then, Alaskan reserves have steadily declined. In 1987 Alaska was estimated to have 13.2 billion barrels of crude oil; by 2002 it had only 4.7 billion barrels. (See Figure 7.1.) Crude oil proved reserves fell by 173 million barrels in Alaska from 2001 to 2002.

The Gulf of Mexico federal offshore areas, which are in U.S. territorial waters, had about 4.4 billion barrels of crude oil proved reserves in 2002 (see Figure 7.1), up 156 million barrels from 2001 to 2002. Improvements in deepwater drilling systems—floating platforms and subsea wells—have allowed the industry to expand into continually deeper Gulf waters in search of crude oil. The Gulf holds much promise for future reserves discoveries.

In 1996 scientists turned their attention to the Permian Basin of west Texas and southeastern New Mexico, where plenty of dry-land potential for crude oil was found, making it one of the most active onshore areas for recent exploration. In 2002 Texas had about five billion barrels of crude oil proved reserves, up seventy-one mil-

FIGURE 7.1

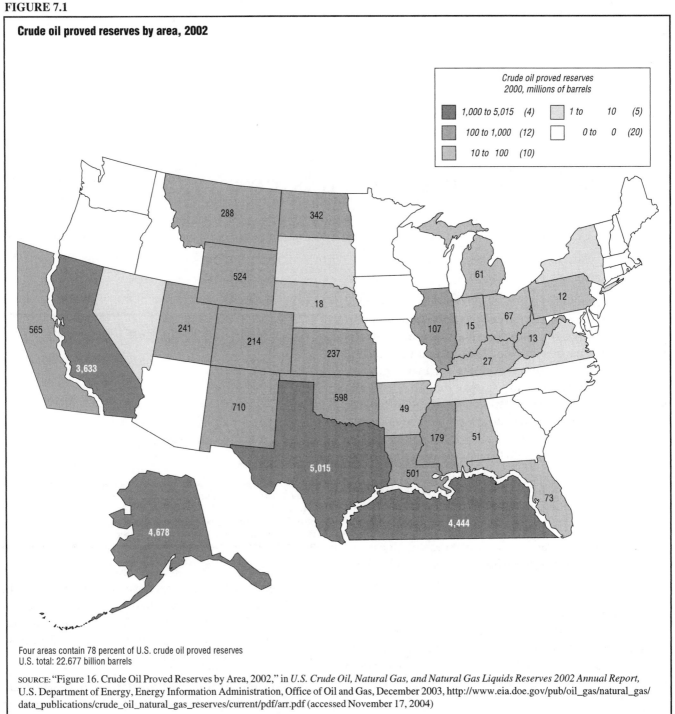

Crude oil proved reserves by area, 2002

Crude oil proved reserves
2000, millions of barrels

1,000 to 5,015 (4)	1 to 10 (5)
100 to 1,000 (12)	0 to 0 (20)
10 to 100 (10)	

288   342
524
565   241   18
3,633   214   61
710   237   107   15   67   12
598   27   13
5,015   49
4,678   179   51
501
4,444   73

Four areas contain 78 percent of U.S. crude oil proved reserves
U.S. total: 22.677 billion barrels

SOURCE: "Figure 16. Crude Oil Proved Reserves by Area, 2002," in *U.S. Crude Oil, Natural Gas, and Natural Gas Liquids Reserves 2002 Annual Report*, U.S. Department of Energy, Energy Information Administration, Office of Oil and Gas, December 2003, http://www.eia.doe.gov/pub/oil_gas/natural_gas/data_publications/crude_oil_natural_gas_reserves/current/pdf/arr.pdf (accessed November 17, 2004)

lion barrels from 2001. Proved reserves in New Mexico dropped by five million barrels from 2001 to 2002.

NATURAL GAS

Figure 7.2 shows that in 2002 the United States had about 187 trillion cubic feet of dry natural gas proved reserves, which reflects an increase of 3.5 billion feet from 2001. The United States also had 196 trillion cubic feet of natural gas liquid proved reserves, which reflects a 2% increase from the volume reported in 2001 in the lower forty-eight states, and a 4% decrease in Alaska.

UNDISCOVERED OIL AND GAS RESOURCES

In addition to those proved resources, other resources are believed to exist based on past geological experience, although they are not yet proved. For 2002 the DOE's Energy Information Administration (EIA) reported in *U.S. Crude Oil, Natural Gas, and Natural Gas Liquids Reserves 2002 Annual Report* that there were an estimated 105 billion barrels of crude oil, 682 trillion cubic feet of dry natural gas, and 8 billion barrels of natural gas liquids as undiscovered resources in the United States.

FIGURE 7.2

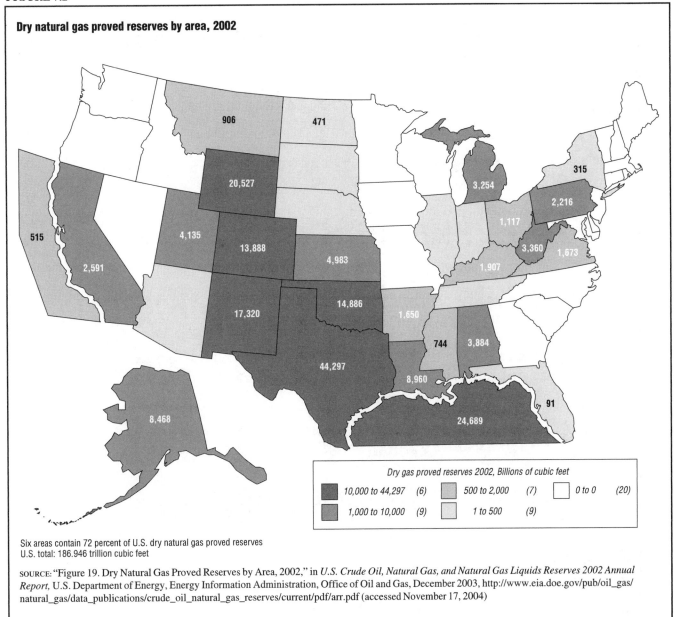

Dry natural gas proved reserves by area, 2002

Dry gas proved reserves 2002, Billions of cubic feet

■	10,000 to 44,297	(6)	▣	500 to 2,000	(7)	□	0 to 0 (20)
▨	1,000 to 10,000	(9)	▢	1 to 500	(9)		

Six areas contain 72 percent of U.S. dry natural gas proved reserves
U.S. total: 186.946 trillion cubic feet

SOURCE: "Figure 19. Dry Natural Gas Proved Reserves by Area, 2002," in *U.S. Crude Oil, Natural Gas, and Natural Gas Liquids Reserves 2002 Annual Report,* U.S. Department of Energy, Energy Information Administration, Office of Oil and Gas, December 2003, http://www.eia.doe.gov/pub/oil_gas/ natural_gas/data_publications/crude_oil_natural_gas_reserves/current/pdf/arr.pdf (accessed November 17, 2004)

Looking for Oil and Gas

Finding oil and gas is usually a two-step process. First, geological and geophysical exploration identifies the areas where oil and gas may most likely be found. Much of this exploration is seismic, which measures the movement of the earth. Then exploratory wells are drilled to determine if oil or gas is present.

Market conditions and technological developments shape exploration for oil and gas. For example, the economic problems of the oil industry in the United States can be seen in the drop in the number of exploratory oil and gas wells completed. Drilling activity for exploratory wells has declined dramatically since the early 1980s. (See Table 7.2.) In 1981 a peak of 91,553 exploratory wells were drilled, with nearly 70% of those successful. In 2003 only 30,151 were attempted, but 85.3% were suc-

cessful. In 1981, 3,970 rotary rigs were in operation; by 2003 only 1,032 were operating. (See Figure 7.3.) Of this number, 872 rigs drilled for natural gas, while 157 drilled for oil. There were 924 onshore rigs and 108 offshore rigs. The average depth of exploratory and development wells has steadily increased, from 3,635 feet in 1949 to 5,958 feet in 2003. Gas wells (averaging 6,224 feet in 2003) are typically deeper than oil wells (5,059 feet).

The Cost to Drill

In 2002 an average well cost about $1,014,200 to drill, or $187.90 per foot. (See Table 7.3.) A gas well generally costs more than an oil well to drill because it is deeper. In 2002, however, the cost of drilling an average gas well ($991,900) was not much higher than that of drilling an oil well ($882,800), because the average cost

TABLE 7.2

Crude oil and natural gas exploratory and development wells, selected years, 1949–2003

Year	Wells drilled				Successful wells (percent)	Footage drilled (thousand feet)				Average depth (feet per well)			
	Crude oil	Natural gas	Dry holes	Total		Crude oil	Natural gas	Dry holes	Total	Crude oil	Natural gas	Dry holes	Total
1949	21,352	3,363	12,597	37,312	66.2	79,428	12,437	43,754	135,619	3,720	3,698	3,473	3,635
1950	23,812	3,439	14,799	42,050	64.8	92,695	13,685	50,977	157,358	3,893	3,979	3,445	3,742
1955	30,432	4,266	20,452	55,150	62.9	121,148	19,930	85,103	226,182	3,981	4,672	4,161	4,101
1960	22,258	5,149	18,212	45,619	60.1	86,568	28,246	77,361	192,176	3,889	5,486	4,248	4,213
1965	18,065	4,482	16,226	38,773	58.2	73,322	24,931	76,629	174,882	4,059	5,562	4,723	4,510
1970	12,968	4,011	11,031	28,010	60.6	56,859	23,623	58,074	138,556	4,385	5,860	5,265	4,943
1971	11,853	3,971	10,309	26,133	60.6	49,109	23,460	54,685	127,253	4,126	5,890	5,305	4,858
1972	11,378	5,440	10,891	27,709	60.7	49,269	30,006	58,556	137,831	4,330	5,516	5,377	4,974
1973	10,167	6,933	10,320	27,420	62.4	44,416	38,045	55,761	138,223	4,369	5,488	5,403	5,041
1974	13,647	7,138	12,116	32,901	63.2	52,025	38,449	62,899	153,374	3,812	5,387	5,191	4,662
1975	16,948	8,127	13,646	38,721	64.8	66,819	44,454	69,220	180,494	3,943	5,470	5,073	4,661
1976	17,688	9,409	13,758	40,855	66.3	68,892	49,113	68,977	186,982	3,895	5,220	5,014	4,577
1977	18,745	12,122	14,985	45,852	67.3	75,451	63,686	76,728	215,866	4,025	5,254	5,120	4,708
1978	19,181	14,413	16,551	50,145	67.0	77,041	75,841	85,788	238,669	4,017	5,262	5,183	4,760
1979	20,851	15,254	16,099	52,204	69.2	82,688	80,468	81,642	244,798	3,966	5,275	5,071	4,689
1980	32,639	17,333	20,638	70,610	70.8	124,350	91,484	98,820	314,654	3,810	5,278	4,788	4,456
1981	43,598	20,166	27,789	91,553	69.6	171,241	107,758	134,113	413,112	3,928	5,344	4,826	4,512
1982	39,199	18,979	26,219	84,397	68.9	148,881	106,627	122,787	378,295	3,798	5,618	4,683	4,482
1983	37,120	14,564	24,153	75,837	68.2	136,078	77,530	104,378	317,986	3,666	5,323	4,322	4,193
1984	42,605	17,127	25,681	85,413	69.9	161,770	90,578	119,044	371,392	3,797	5,289	4,635	4,348
1985	35,118	14,168	21,056	70,342	70.1	137,366	75,862	99,816	313,045	3,912	5,355	4,740	4,450
1986	19,097	8,516	12,678	40,291	68.5	76,622	44,727	60,507	181,856	4,012	5,252	4,773	4,514
1987	16,164	8,055	11,112	35,331	68.5	66,317	42,479	53,382	162,178	4,103	5,274	4,804	4,590
1988	13,636	8,555	10,041	32,232	68.8	58,660	45,320	52,375	156,354	4,302	5,297	5,216	4,851
1989	10,204	9,539	8,188	27,931	70.7	43,287	49,169	41,983	134,439	4,242	5,154	5,127	4,813
1990	12,198	11,044	8,313	31,555	73.7	54,480	55,869	43,352	153,701	4,466	5,059	5,215	4,871
1991	11,770	9,526	7,596	28,892	73.7	54,283	49,737	39,001	143,021	4,612	5,221	5,134	4,950
1992	8,757	8,209	6,118	23,084	73.5	44,183	45,728	31,213	121,124	5,045	5,571	5,102	5,247
1993	8,407	10,017	6,328	24,752	74.4	42,895	59,720	32,503	135,118	5,102	5,962	5,136	5,459
1994	6,721	9,538	5,307	21,566	75.4	36,090	59,412	29,306	124,809	5,370	6,229	5,522	5,787
1995	7,627	8,354	5,075	21,056	75.9	38,024	51,415	28,393	117,832	4,985	6,154	5,595	5,596
1996	8,314	9,302	5,282	22,898	76.9	40,849	58,062	30,133	129,045	4,913	6,242	5,705	5,636
1997	10,436	11,327	5,702	27,465	79.2	52,098	70,477	34,086	156,661	4,992	6,222	5,978	5,704
1998	7,064	R11,308	4,840	R23,212	R79.1	R37,576	R74,194	R31,683	143,454	5,321	6,672	6,512	6,224
1999E	4,176	10,877	3,364	18,417	81.7	R19,793	R58,242	R21,375	99,410	4,723	5,393	6,250	5,398
2000E	7,358	16,455	4,025	27,838	85.5	34,691	R83,091	R23,610	141,392	4,698	5,083	5,762	5,079
2001E	8,060	22,083	R4,084	R34,227	R88.1	42,504	R120,307	R27,156	189,967	R5,274	R5,448	R6,649	R5,550
2002E	R6,058	15,947	R3,531	R25,536	86.2	R27,375	R91,190	R19,744	R138,310	R4,519	R5,718	R5,592	R5,416
2003E	5,694	20,011	4,446	30,151	85.3	28,808	124,543	26,286	179,637	5,059	6,224	5,912	5,958

TABLE 7.2

Crude oil and natural gas exploratory and development wells, selected years, 1949–2003 [CONTINUED]

R = Revised.
E = Estimate.
Notes: Data are for all wells. Service wells, stratigraphic tests, and core tests are excluded. For 1949–1959, data represent wells completed in a given year. For 1960–1969, data are for well completion reports received by the American Petroleum Institute during the reporting year. For 1970 forward, the data represent wells completed in a given year. The as-received well completion data for recent years are incomplete due to delays in the reporting of wells drilled. The Energy Information administration (EIA) therefore statistically imputes the missing data to provide estimates of total well completions and footage where necessary. Totals may not equal sum of components due to independent rounding. Average depth may not equal average of components due to independent rounding.
Web Pages: For data not shown for 1951–1969, see http://www.eia.doe.gov/emeu/aer/resource.html. For related information, see http://www.eia.doe.gov/oil_gas/petroleum/info_glance/petroleum.html.

SOURCE: "Table 4.4. Crude Oil and Natural Gas Exploratory and Development Wells, Selected Years, 1949–2003," in *Annual Energy Review 2003*, U.S. Department of Energy, Energy Information Administration, Office of Energy Markets and End Use, September 7, 2004, http://www.eia.doe.gov/emeu/aer/pdf/aer.pdf (accessed September 28, 2004)

Energy

FIGURE 7.3

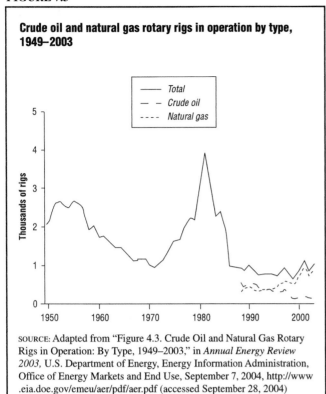

Crude oil and natural gas rotary rigs in operation by type, 1949–2003

SOURCE: Adapted from "Figure 4.3. Crude Oil and Natural Gas Rotary Rigs in Operation: By Type, 1949–2003," in *Annual Energy Review 2003*, U.S. Department of Energy, Energy Information Administration, Office of Energy Markets and End Use, September 7, 2004, http://www .eia.doe.gov/emeu/aer/pdf/aer.pdf (accessed September 28, 2004)

per foot of drilling an oil well ($194.55) was higher than that of drilling a foot of gas well ($175.78). Although drilling costs have fluctuated in recent years, it costs considerably more to drill a well today than it did in the 1960s and early 1970s, not only because of inflation but also because wells must now be drilled deeper. Even in as short a period as 1996 to 2002, the cost per well nearly doubled, costing $485,600 more per well.

Spending on Exploration and Development

The estimated expenditures on exploration for, and development of, oil and gas by major U.S. energy-producing companies peaked at $76.8 billion in 2000. Another high point was in 1984, when companies spent $65.3 billion. (See Table 7.4.) In 2002 U.S. energy companies spent $63.2 billion in 2002 on oil and gas exploration around the world, with about half of that, $31.8 billion, spent for exploration in the United States.

The chance of finding a major oil or gas field has become increasingly small. Oil explorers are spending more of their time looking for less oil. This does not necessarily imply that the nation will soon be facing a shortage of gas or oil; reserves are still considerable, especially of natural gas. It does show, however, that the oil and gas industries are mature, or highly developed.

Drilling in the Arctic National Wildlife Refuge (ANWR)

Much controversy has developed over opening the Arctic National Wildlife Refuge (ANWR) in Alaska to oil

drilling. Prudhoe Bay, directly to the west of the refuge, supplies about 60% of Alaska's oil and 20% of the country's domestic oil, although production is dropping steadily as the oil is used up. In 1980 Congress passed the Alaska National Interest Lands Conservation Act (PL 96-487), which set aside more than 104 million acres for parks, refuges, and wilderness areas, including ANWR (18 million acres), but the conservation act did not include the coastal plain.

The U.S. Department of the Interior (DOI), in a 1987 report to Congress, recommended that the 1.5-million-acre coastal plain be opened for exploration and extraction. Alaskan corporations supported the proposal in hopes of sharing in the proceeds. Environmentalists strongly opposed the plan because of the destruction that drilling could cause to native wildlife, such as caribou, polar and grizzly bears, musk ox, wolves, Arctic foxes, and millions of nesting birds.

The DOI's Fish and Wildlife Service estimated that 3.2 billion barrels of recoverable oil exist in the contested area. They also estimated that there was a 46% chance of recovering the oil, a high figure by industry standards. Considering the general decline in American production, this could account for a significant portion of American petroleum production in this century. Vast quantities of natural gas are also likely to be found in the area.

The report further stated, however, that there could be a major effect on the migratory caribou herds, which number about 180,000 animals. While environmentalists estimated that 20–40% of the animals would be threatened, DOI officials claimed that the caribou would change their migratory habits.

If major oil reserves are found, oil companies might operate on the coastal plain for several decades. As of 2004 no legislation had yet been passed to allow or disallow drilling in ANWR.

COAL

The EIA estimated U.S. coal reserves at 497.7 billion short tons on January 1, 2003. (See Table 7.5.) About 42% of this, 211.5 billion short tons, is underground bituminous coal. Montana, Illinois, and Wyoming have the largest coal reserves.

In addition to untapped coal reserves, large stockpiles of coal are maintained by coal producers, distributors, and major consumers (such as electric utility companies and industrial plants) to compensate for possible interruptions in supply. Although there is little seasonal change in demand for coal, supply can vary owing to factors such as coal miners' strikes and bad weather. According to the EIA's *Annual Energy Review 2003* (2004), coal stockpiles totaled 163.8 million short tons in 2003. Electric utilities

TABLE 7.3

Costs of crude oil and natural gas wells drilled, 1960–2002

Year	Costs per well (thousand dollars)					Costs per foot (dollars)				
	Crude oil Nominal	Natural gas Nominal	Dry holes Nominal	All Nominal	All Real*	Crude oil Nominal	Natural gas Nominal	Dry holes Nominal	All Nominal	All Real*
1960	52.2	102.7	44.0	54.9	R261.1	13.22	18.57	10.56	13.01	R61.83
1961	51.3	94.7	45.2	54.5	R256.2	13.11	17.65	10.56	12.85	R60.39
1962	54.2	97.1	50.8	58.6	R271.8	13.41	18.10	11.20	13.31	R61.71
1963	51.8	92.4	48.2	55.0	R252.4	13.20	17.19	10.58	12.69	R58.22
1964	50.6	104.8	48.5	55.8	R252.2	13.12	18.57	10.64	12.86	R58.11
1965	56.6	101.9	53.1	60.6	R269.1	13.94	18.35	11.21	13.44	R59.64
1966	62.2	133.8	56.9	68.4	R295.1	15.04	21.75	12.34	14.95	R64.51
1967	66.6	141.0	61.5	72.9	R305.1	16.61	23.05	12.87	15.97	R66.84
1968	79.1	148.5	66.2	81.5	R327.0	18.63	24.05	12.88	16.83	R67.56
1969	86.5	154.3	70.2	88.6	R338.7	19.28	25.58	13.23	17.56	R67.15
1970	86.7	160.7	80.9	94.9	R344.6	19.29	26.75	15.21	18.84	R68.42
1971	78.4	166.6	86.8	94.7	R327.6	18.41	27.70	16.02	19.03	R65.82
1972	93.5	157.8	94.9	106.4	R352.8	20.77	27.78	17.28	20.76	R68.82
1973	103.8	155.3	105.8	117.2	R367.8	22.54	27.46	19.22	22.50	R70.65
1974	110.2	189.2	141.7	138.7	R399.5	27.82	34.11	26.76	28.93	R83.31
1975	138.6	262.0	177.2	177.8	R467.9	34.17	46.23	33.86	36.99	R97.34
1976	151.1	270.4	190.3	191.6	R476.7	37.35	49.78	36.94	40.46	R100.66
1977	170.0	313.5	230.2	227.2	R531.4	41.16	57.57	43.49	46.81	R109.49
1978	208.0	374.2	281.7	280.0	R611.8	49.72	68.37	52.55	56.63	R123.76
1979	243.1	443.1	339.6	331.4	R668.8	58.29	80.66	64.60	67.70	R136.64
1980	272.1	536.4	376.5	367.7	R680.4	66.36	95.16	73.70	77.02	R142.52
1981	336.3	698.6	464.0	453.7	R767.4	80.40	122.17	90.03	94.30	R159.51
1982	347.4	864.3	515.4	514.4	R820.0	86.34	146.20	104.09	108.73	R173.34
1983	283.8	608.1	366.5	371.7	R570.1	72.65	108.37	79.10	83.34	R127.81
1984	262.1	489.8	329.2	326.5	R482.5	66.32	88.80	67.18	71.90	R106.27
1985	270.4	508.7	372.3	349.4	R501.2	66.78	93.09	73.69	75.35	R108.09
1986	284.9	522.9	389.2	364.6	R511.7	68.35	93.02	76.53	76.88	R107.90
1987	246.0	380.4	259.1	279.6	R382.0	58.35	69.55	51.05	58.71	R80.21
1988	279.4	460.3	366.4	354.7	R468.6	62.28	84.65	66.96	70.23	R92.78
1989	282.3	457.8	355.4	362.2	R461.1	64.92	86.86	67.61	73.55	R93.63
1990	321.8	471.3	367.5	383.6	R470.2	69.17	90.73	67.49	76.07	R93.23
1991	346.9	506.6	441.2	421.5	R499.1	73.75	93.10	83.05	82.64	R97.86
1992	362.3	426.1	357.6	382.6	R442.9	69.50	72.83	67.82	70.27	R81.35
1993	356.6	521.2	387.7	426.8	R482.9	67.52	83.15	72.56	75.30	R85.20
1994	409.5	535.1	491.5	483.2	R535.4	70.57	81.90	86.60	79.49	R88.07
1995	415.8	629.7	481.2	513.4	R557.4	78.09	95.97	84.60	87.22	R94.70
1996	341.0	616.0	541.0	496.1	R528.6	70.60	98.67	95.74	88.92	R94.74
1997	445.6	728.6	655.6	603.9	R632.9	90.48	117.55	115.09	107.83	R113.01
1998	566.0	815.6	973.2	769.1	R797.2	108.88	127.94	157.79	128.97	R133.69
1999	783.0	798.4	1,115.5	856.1	R874.8	156.45	138.42	182.99	152.02	R155.33
2000	593.4	756.9	1,075.4	754.6	R754.6	125.96	138.39	181.83	142.16	R142.16
2001	729.1	896.5	1,620.4	943.2	R921.3	153.72	172.05	271.63	181.94	R177.72
2002	882.8	991.9	1,673.4	1,054.2	1,014.2	194.55	175.78	284.17	195.31	187.90

TABLE 7.3

Costs of crude oil and natural gas wells drilled, 1960–2002 [CONTINUED]

*In chained (2000) dollars, calculated by using gross domestic product implicit price deflators.

R = Revised.

Notes: The information reported for 1965 and prior years is not strictly comparable to that in more recent surveys. Average cost is the arithmetic mean and includes all costs for drilling and equipping wells and for surface-producing facilities. Wells drilled include exploratory and development wells; excludes service wells, stratigraphic tests, and core tests.

Web Page: For related information, see http://api-ec.api.org/newsplashpage/index.cfm.

SOURCE: "Table 4.7. Costs of Crude Oil and Natural Gas Wells Drilled, 1960–2002," in *Annual Energy Review 2003*, U.S. Department of Energy, Energy Information Administration, Office of Energy Markets and End Use, September 7, 2004, http://www.eia.doe.gov/emeu/aer/pdf/aer.pdf (accessed September 28, 2004)

TABLE 7.4

Major U.S. energy companies' expenditures for crude oil and natural gas exploration and development by region, 1974–2002

(Billion dollars[1])

Year	United States			Foreign								Total
	Onshore	Offshore	Total	Canada	OECD Europe[2]	Eastern Europe and Former U.S.S.R.	Africa	Middle East	Other Eastern Hemisphere[3]	Other Western Hemisphere[4]	Total	
1974	NA	NA	8.7	NA	NA	—	NA	NA	NA	NA	3.8	12.5
1975	NA	NA	7.8	NA	NA	—	NA	NA	NA	NA	5.3	13.1
1976	NA	NA	9.5	NA	NA	—	NA	NA	NA	NA	5.2	14.7
1977	6.7	4.0	10.7	1.5	2.5	—	0.7	0.2	0.3	0.4	5.6	16.3
1978	7.5	4.3	11.8	1.6	2.6	—	0.8	0.3	0.4	0.6	6.4	18.2
1979	13.0	8.3	21.3	2.3	3.0	—	0.8	0.2	0.5	0.8	7.8	29.1
1980	16.8	9.4	26.2	3.1	4.3	—	1.4	0.2	0.8	1.0	11.0	37.2
1981	19.9	13.0	33.0	1.8	5.0	—	2.1	0.3	1.9	1.3	12.4	45.4
1982	27.2	11.9	39.1	1.9	6.3	—	2.1	0.4	2.4	1.1	14.2	53.3
1983	16.0	11.1	27.1	1.6	4.3	—	1.7	0.5	2.0	0.6	10.7	37.7
1984	32.1	16.0	48.1	5.4	5.5	—	3.4	0.5	2.0	0.5	17.3	65.3
1985	20.0	8.5	28.5	1.9	3.7	—	1.6	0.9	1.3	0.7	10.1	38.6
1986	12.5	4.9	17.4	1.1	3.2	—	1.1	0.3	1.2	0.6	7.5	24.9
1987	9.7	4.5	14.3	1.9	3.0	—	0.8	0.4	2.8	0.5	9.2	23.5
1988	12.9	8.1	21.0	5.4	4.3	—	0.8	0.4	1.4	0.7	13.0	34.1
1989	9.0	6.0	15.0	6.3	3.5	—	1.0	0.4	2.3	0.6	14.1	29.1
1990	10.2	4.9	15.1	1.8	6.6	—	1.4	0.6	2.4	0.7	13.6	28.7
1991	9.6	4.6	14.2	1.7	6.8	—	1.5	0.5	2.4	0.7	13.7	27.9
1992	7.3	3.0	10.3	1.1	6.8	—	1.4	0.6	2.4	0.6	12.9	23.2
1993	7.2	3.7	10.9	1.6	5.5	0.3	1.5	0.7	2.5	0.6	12.5	23.5
1994	7.8	4.8	12.6	1.8	4.4	0.3	1.4	0.4	2.8	0.7	11.9	24.5
1995	7.7	4.7	12.4	1.9	5.2	0.4	2.0	0.4	2.4	0.9	13.2	25.6
1996	7.9	6.7	14.6	1.6	5.6	0.5	2.8	0.5	4.1	1.6	16.6	31.3
1997	13.0	8.8	21.8	2.0	7.1	0.6	3.0	0.6	3.0	1.6	17.9	39.8
1998	13.5	11.0	24.4	4.8	8.6	1.3	3.1	0.9	3.9	3.7	26.4	50.8
1999	6.6	6.9	13.5	2.1	4.1	0.6	3.1	0.4	3.4	3.8	17.5	31.0
2000	27.1	21.0	48.0	4.9	7.5	0.9	2.7	0.6	6.8	5.4	28.8	76.8
2001	24.2	9.6	33.9	15.3	5.4	0.9	5.5	0.7	5.0	3.1	35.9	69.8
2002	22.3	9.5	31.8	6.7	9.8	1.3	5.1	0.8	6.2	1.6	31.4	63.2

[1]Nominal dollars.
[2]The European members of the Organization for Economic Cooperation and Development (OECD) are Austria, Belgium, Denmark, Finland, France, Germany, Greece, Iceland, Ireland, Italy, Luxembourg, the Netherlands, Norway, Portugal, Spain, Sweden, Switzerland, Turkey, and the United Kingdom, and, for 1997 forward, Czech Republic, Hungary, and Poland.
[3]This region includes areas that are eastward of the Greenwich prime meridian to 180° longitude and that are not included in other domestic or foreign classifications.
[4]This region includes areas that are westward of the Greenwich prime meridian to 180° longitude and that are not included in other domestic or foreign classifications.
NA=Not available.
— = Not applicable.
Notes: "Major U.S. Energy Companies" are the top publicly-owned, U.S.-based crude oil and natural gas producers and petroleum refiners that form the Financial Reporting System (FRS). Totals may not equal sum of components due to independent rounding.
Web Page: For related information, see http://www.eia.doe.gov/emeu/finance.

SOURCE: "Table 4.9. Major U.S. Energy Companies' Expenditures for Crude Oil and Natural Gas Exploration and Development by Region, 1974–2002 (Billion Dollars)," in Annual Energy Review 2003, U.S. Department of Energy, Energy Information Administration, Office of Energy Markets and End Use, September 7, 2004, http://www.eia.doe.gov/emeu/aer/pdf/aer.pdf (accessed September 28, 2004)

TABLE 7.5

Coal demonstrated reserve base, January 1, 2003

(Billion short tons)

Region and state	Anthracite	Bituminous coal		Subbituminous coal		Lignite	Total		
		Underground	Surface	Underground	Surface	Surface[1]	Underground	Surface	Total
Appalachian	**7.3**	**71.9**	**23.3**	**0.0**	**0.0**	**1.1**	**75.9**	**27.7**	**103.6**
Alabama	0.0	1.1	2.1	0.0	0.0	1.1	1.1	3.2	4.3
Kentucky, Eastern	0.0	1.5	9.5	0.0	0.0	0.0	1.5	9.5	11.0
Ohio	0.0	17.6	5.8	0.0	0.0	0.0	17.6	5.8	23.4
Pennsylvania	7.2	19.7	0.9	0.0	0.0	0.0	23.5	4.3	27.8
Virginia	0.1	1.1	0.6	0.0	0.0	0.0	1.2	0.6	1.8
West Virginia	0.0	29.7	4.0	0.0	0.0	0.0	29.7	4.0	33.7
Other[2]	0.0	1.1	0.3	0.0	0.0	0.0	1.1	0.3	1.5
Interior	**0.1**	**117.6**	**27.4**	**0.0**	**0.0**	**13.0**	**117.7**	**40.5**	**158.2**
Illinois	0.0	88.1	16.6	0.0	0.0	0.0	88.1	16.6	104.6
Indiana	0.0	8.8	0.8	0.0	0.0	0.0	8.8	0.8	9.6
Iowa	0.0	1.7	0.5	0.0	0.0	0.0	1.7	0.5	2.2
Kentucky, Western	0.0	16.0	3.6	0.0	0.0	0.0	16.0	3.6	19.6
Missouri	0.0	1.5	4.5	0.0	0.0	0.0	1.5	4.5	6.0
Oklahoma	0.0	1.2	0.3	0.0	0.0	0.0	1.2	0.3	1.6
Texas	0.0	0.0	0.0	0.0	0.0	12.6	0.0	12.6	12.6
Other[3]	0.1	0.3	1.1	0.0	0.0	0.5	0.4	1.6	2.0
Western	**(s)**	**22.0**	**2.4**	**121.3**	**60.7**	**29.5**	**143.4**	**92.6**	**235.9**
Alaska	0.0	0.6	0.1	4.8	0.6	(s)	5.4	0.7	6.1
Colorado	(s)	7.9	0.6	3.8	0.0	4.2	11.7	4.8	16.4
Montana	0.0	1.4	0.0	69.6	32.7	15.8	71.0	48.4	119.4
New Mexico	(s)	2.7	0.9	3.5	5.1	0.0	6.2	6.1	12.2
North Dakota	0.0	0.0	0.0	0.0	0.0	9.2	0.0	9.2	9.2
Utah	0.0	5.3	0.3	0.0	0.0	0.0	5.3	0.3	5.5
Washington	0.0	0.3	0.0	1.0	(s)	(s)	1.3	0.0	1.4
Wyoming	0.0	3.8	0.5	38.7	22.3	0.0	42.5	22.8	65.3
Other[4]	0.0	0.0	0.0	(s)	(s)	0.4	0.0	0.4	0.4
U.S. total	**7.5**	**211.5**	**53.1**	**121.3**	**60.7**	**43.6**	**336.9**	**160.8**	**497.7**
States East of the Mississippi River	7.3	184.9	44.3	0.0	0.0	1.1	188.8	48.8	237.6
States West of the Mississippi River	0.1	26.6	8.7	121.3	60.7	42.5	148.1	112.0	260.1

[1]Lignite resources are not mined underground in the United States.
[2]Georgia, Maryland, North Carolina, and Tennessee.
[3]Arkansas, Kansas, Louisiana, and Michigan.
[4]Arizona, Idaho, Oregon, and South Dakota.
(s)=Less than 0.05 billion short tons.
Notes: See *U.S. Coal Reserves: 1997 Update* on the Web Page for a description of the methodology used to produce these data. Data represent known measured and indicated coal resources meeting minimum seam and depth criteria, in the ground as of January 1, 2003. These coal resources are not totally recoverable. Net recoverability with current mining technologies ranges from 0 percent (in far northern Alaska) to more than 90 percent. Fifty-four percent of the demonstrated reserve base of coal in the United States is estimated to be recoverable.
Totals may not equal sum of components due to independent rounding.
Web Page: For related information, see http://www.eia.doe.gov/fuelcoal.html.

SOURCE: "Table 4.11. Coal Demonstrated Reserve Base, January 1, 2003 (Billion Short Tons)," in *Annual Energy Review 2003,* U.S. Department of Energy, Energy Information Administration, Office of Energy Markets and End Use, September 7, 2004, http://www.eia.doe.gov/emeu/aer/pdf/aer.pdf (accessed September 28, 2004)

---

held slightly more than 74% of this coal, and coal producers and distributors stocked another 22%.

URANIUM

The United States possesses enough uranium to fuel existing nuclear reactors for more than forty years. In 2003, according to the EIA's "U.S. Uranium Reserves by Forward-Cost Category, Year-End 1993-2003," uranium reserves totaled 1.4 billion pounds of uranium oxide, mostly in Wyoming and New Mexico. Exploration for uranium has reflected that energy markets are moving away from nuclear energy. The number of uranium mills producing uranium concentrate dropped from eleven at the end of 1997 to three at the end of September 2004 (EIA, *Domestic Uranium Pro-*

duction Report, "Number of Uranium Mills and Plants Producing Uranium Concentrate," http://eia.doe.gov/cneaf/nuclear/dupr/qupd.html [accessed January 10, 2005]).

INTERNATIONAL RESERVES

Crude Oil

Between 57% and 65% of the estimated world crude oil reserves of approximately one trillion barrels (as of January 1, 2003) are located in the Middle East. (See Table 7.6. This graphic gives two different sets of estimates—one from Pennwell Publishing Company's *Oil & Gas Journal* and one from Gulf Publishing Company's *World Oil.*) Saudi Arabia, Iraq, United Arab

TABLE 7.6

Region and country	Crude oil (billion barrels)		Natural gas (trillion cubic feet)	
	Oil & gas Journal	World oil	Oil & gas Journal	World oil
North America	**215.3**	**45.4**	**255.8**	**262.1**
Canada	[1]180.0	5.5	60.1	60.1
Mexico	12.6	17.2	8.8	15.0
United States	22.7	22.7	186.9	186.9
Central and South America	**98.6**	**75.9**	**250.1**	**244.4**
Argentina	2.9	2.8	27.0	23.4
Bolivia	0.4	0.9	24.0	28.1
Brazil	8.3	9.8	8.1	8.4
Colombia	1.8	1.6	4.5	4.2
Ecuador	4.6	4.6	0.3	0.3
Peru	0.3	1.0	8.7	8.6
Trinidad and Tobago	0.7	1.0	23.5	20.3
Venezuela	77.8	53.1	148.0	149.2
Other	1.6	1.0	6.1	1.8
Western Europe	**18.3**	**17.0**	**191.6**	**175.7**
Denmark	1.3	1.8	3.0	4.2
Germany	0.3	0.3	11.3	8.5
Italy	0.6	0.7	8.0	7.9
Netherlands	0.1	0.1	62.0	55.3
Norway	10.3	9.0	77.3	74.7
United Kingdom	4.7	4.5	24.6	22.2
Other	0.9	0.6	5.4	2.9
Eastern Europe and Former U.S.S.R.	**79.2**	**81.9**	**1,964.2**	**2,047.0**
Hungary	0.1	0.1	1.2	2.2
Kazakhstan	9.0	NA	65.0	NA
Romania	1.0	1.1	3.6	4.2
Russia	60.0	58.8	1,680.0	1,700.0
Other[3]	9.1	21.9	214.4	340.6
Middle East	**685.6**	**669.8**	**1,979.7**	**2,517.0**
Bahrain	0.1	NA	3.3	NA
Iran	89.7	100.1	812.3	913.6
Iraq	112.5	115.0	109.8	112.6
Kuwait[2]	96.5	98.9	52.7	56.6
Oman	5.5	5.7	29.3	31.0
Qatar	15.2	19.6	508.5	916.0
Saudi Arabia[2]	261.8	261.8	224.7	234.6
Syria	2.5	2.3	8.5	18.0
United Arab Emirates	97.8	63.0	212.1	204.1
Yemen	4.0	2.9	16.9	17.0
Other	(s)	0.7	1.6	13.5
Africa	**77.4**	**96.3**	**418.2**	**438.9**
Algeria	9.2	13.0	159.7	170.0
Angola	5.4	8.9	1.6	4.0
Cameroon	0.4	NA	3.9	NA
Congo	1.5	1.5	3.2	4.2
Egypt	3.7	3.5	58.5	5.9
Libya	29.5	30.0	46.4	46.0
Nigeria	24.0	32.0	124.0	178.5
Tunisia	0.3	0.5	2.8	2.7
Other	3.4	6.8	18.1	27.7

Region and country	Crude oil (billion barrels)		Natural gas (trillion cubic feet)	
	Oil & gas Journal	World oil	Oil & gas Journal	World oil
Asia and Oceania	**38.7**	**48.5**	**445.4**	**441.7**
Australia	3.5	3.7	90.0	85.0
Brunei	1.4	1.1	13.8	8.3
China	18.3	23.7	53.3	46.7
India	5.4	4.6	26.9	23.6
Indonesia	5.0	5.9	92.5	73.5
Japan	0.1	NA	1.4	NA
Malaysia	3.0	4.3	75.0	88.0
New Zealand	0.2	0.1	3.1	2.6
Pakistan	0.3	0.3	26.4	26.4
Papua New Guinea	0.2	0.4	12.2	13.5
Thailand	0.6	0.5	13.3	12.9
Other	0.9	3.8	37.4	61.4
World	**1,213.1**	**1,034.7**	**5,504.9**	**6,126.6**

[1]Includes 5.2 billion barrels of conventional crude oil and condensate reserves and 174.8 billion barrels of bitumen that is contained in Alberta's oil sands.
[2]Data for Kuwait and Saudi Arabia include one-half of the reserves in the Neutral Zone between Kuwait and Saudi Arabia.
[3]Albania, Azerbaijan, Belarus, Bulgaria, Czech Republic, Georgia, Kyrgyzstan, Lithuania, Poland, Slovakia, Tajikistan, Turkmenistan, Ukraine, Uzbekistan.
NA=Not available.
(s)=Less than 0.05 billion barrels.
Notes: All reserve figures except those for the former U.S.S.R. and natural gas reserves in Canada are proved reserves recoverable with present technology and prices at the time of estimation. Former U.S.S.R. and Canadian natural gas figures include proved, and some probable reserves. Totals may not equal sum of components due to independent rounding.
Web Page: For related information, see http://www.eia.doe.gov/international.

SOURCE: "Table 11.4. World Crude Oil and Natural Gas Reserves, January 1, 2003," in *Annual Energy Review 2003,* U.S. Department of Energy, Energy Information Administration, Office of Energy Markets and End Use, September 7, 2004, http://www.eia.doe.gov/emeu/aer/pdf/aer.pdf (accessed September 28, 2004)

sources for oil. Companies from many nations have turned their attention to areas in the former Soviet Union, including the Caspian Sea, Azerbaijan, and Kazakhstan. Geologists from the United States Geological Survey reported that many billions of barrels of oil may be found there in *World Petroleum Assessment 2000* ("World Undiscovered Assessment Results Summary," U.S. Geological Survey Digital Data Series 60, http://pubs. usgs.gov/dds/dds-060 [accessed January 10, 2005]). The report also concluded that significant amounts of oil may lie in the northeast Greenland Shelf region, the Niger and Congo delta areas, and offshore Suriname in South America.

Energy experts with the EIA predict increased offshore exploration, including in deep water, in *International Energy Outlook 2004,* "World Oil Markets." Offshore production is expected to increase near Algeria, Nigeria, and Venezuela, as well as in the North Sea and offshore areas of West Africa, Brazil, Colombia, Mexico, and Canada. Deep-sea exploration is extremely expensive, but if oil prices rise, the oil industry may launch more deepwater explorations.

Emirates, Kuwait, Iran, Canada, Venezuela, and Russia have the largest reserves. With an estimated twenty-three billion barrels of reserves, or only about 2% of the world's oil reserves total, the United States can no longer depend on its own oil reserves unless it drastically lowers consumption.

Oil industry experts say that even though the world currently has sufficient supplies of oil, demand will rise in the future as economies such as those of China and India expand. This knowledge drives oil companies to seek new

FIGURE 7.4

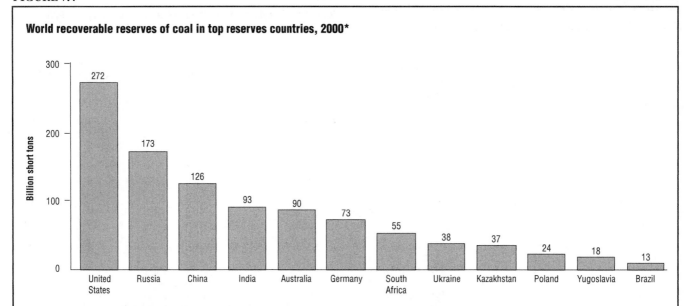

World recoverable reserves of coal in top reserves countries, 2000*

*Recoverable reserves are as of December 31, 2000, except for U.S. recoverable reserves, which are as of December 31, 2002.

SOURCE: Adapted from "Figure 11.13. World Recoverable Reserves of Coal: Top Reserves Countries," in *Annual Energy Review 2003,* U.S. Department of Energy, Energy Information Administration, Office of Energy Markets and End Use, September 7, 2004, http://www.eia.doe.gov/emeu/aer/pdf/aer.pdf (accessed September 28, 2004)

Natural Gas

Russia and the Middle East have approximately two-thirds of the world's estimated 5,505–6,127 trillion cubic feet (as of 2003) of natural gas reserves. (See Table 7.6.) Russia has more than twice as much natural gas in reserve as any other country, while Iran and Qatar possess the largest natural gas reserves in the Middle East. Large reserves (more than one hundred trillion cubic feet) are also located in United Arab Emirates, Saudi Arabia, the United States, Algeria, Venezuela, Nigeria, Iraq, and other areas of the former Soviet Union.

Coal

In 2000 worldwide recoverable reserves of coal were estimated at about 1.1 trillion short tons. The three countries with the most plentiful coal reserves are the United States (272 billion short tons), Russia (173 billion short tons), and China (126 billion short tons). (See Figure 7.4. Note that the data for the United States are as of December 31, 2002, while those for other countries are as of December 31, 2000, the latest available.)

Uranium

The world's supply of uranium is much larger than the capacity for disposing of it, should it be used as fuel. The countries with the largest known uranium reserves as of 2001, according to the Organization for Economic Cooperation and Development and the International Atomic Energy Agency, are Australia, Kazakhstan, Canada, South Africa, Namibia, Brazil, the Russian Federation, the United States, and Uzbekistan.

CHAPTER 8

ELECTRICITY

Since 1879, when Thomas Edison flipped the first switch to light Menlo Park, New Jersey, the use of electrical power has become nearly universal in the United States.

ELECTRICITY DEFINED

Electricity is a form of energy resulting from the movement of charged particles, such as electrons (negatively charged subatomic particles) and protons (positively charged subatomic particles). Static electricity is caused by friction, when one material rubs against another and transfers charged particles. The zap you might feel and the spark you might see when you drag your feet along the carpet and then touch a metal doorknob is static electricity—electrons being transferred between you and the doorknob.

Electric current is the flow of electric charge; it is measured in amperes (amps). Electrical power is the rate at which energy is transferred by electric current. A watt is the standard measure of electrical power, named after the Scottish engineer James Watt. The term "wattage" refers to the amount of electrical power required to operate a particular appliance or device. A kilowatt is a unit of electrical power equal to one thousand watts, while a kilowatt-hour is a unit of electrical work equal to that done by one kilowatt acting for one hour.

Electrical Capacity

The generating capacity of an electrical plant, measured in watts, indicates its ability to produce electrical power. A one-thousand-kilowatt generator running at full capacity for one hour supplies one thousand kilowatt-hours of power. That generator operating continuously for an entire year could produce 8.76 million kilowatt-hours of electricity (1,000 kilowatts x 24 hours/day x 365 days a year). However, no generator can operate at 100% capacity during an entire year because of downtime for routine maintenance, outages, and legal restrictions. On average,

about one-fourth of the potential generating capacity of an electrical plant is not available at any given time.

Electricity demands vary daily and seasonally, so the continuous operation of electrical generators is not necessary. Utilities depend on steam, nuclear, and large hydroelectric plants to meet routine demand. Auxiliary gas, turbine, internal combustion, and smaller hydroelectric plants are normally used during short periods of high demand.

An Electric Power System

An electric power system has several components. Figure 8.1 illustrates a simple electric system. Generating units (power plants) produce electricity, transmission lines carry electricity over long distances, and distribution lines deliver the electricity to customers. Substations connect the pieces of the system together, while energy control centers coordinate the operation of all the components.

U.S. ELECTRICITY USAGE

In 2003, net generation of electricity totaled 3.8 trillion kilowatt-hours. Table 8.1 shows that electric utility retail sales (electricity use) in the United States has increased almost every year since 1949.

In the United States coal has been and continues to be the largest raw source for electricity production, accounting for slightly more than half of the electricity (about 52%) generated in 2003. (See Figure 8.2 and Figure 8.3.) Nuclear power was the second largest source of electricity (approximately 20%), followed by natural gas (14%), hydroelectric power (a renewable energy source, 7%), and petroleum (3%). Very little electricity (2%) was generated in the United States by all other renewable sources combined, such as geothermal, solar, and wind power.

The Energy Information Administration (EIA) of the U.S. Department of Energy (DOE), in its *Annual Energy Review 2003* (2004), noted that coal accounted for the

FIGURE 8.1

A simple electric system

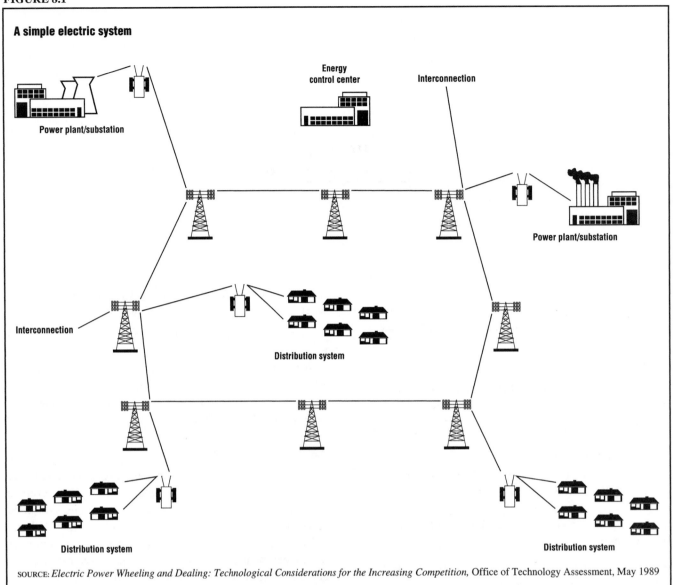

SOURCE: *Electric Power Wheeling and Dealing: Technological Considerations for the Increasing Competition,* Office of Technology Assessment, May 1989

generation of approximately 2 trillion kilowatt-hours of electricity in 2003, while natural gas contributed 629 billion kilowatt-hours and petroleum only 118 billion kilowatt-hours in all sectors. Nuclear power accounted for 764 billion kilowatt-hours, and hydroelectric generation totaled 275 billion kilowatt-hours. All other renewable energy sources, including geothermal, wood, municipal waste, wind, and solar energy, produced only 84 billion kilowatt-hours in 2003.

From 1949 through the early 1990s the industrial sector was the largest consumer of electricity in the United States, but since then sales to the residential sector have generally been higher because of changing economic factors. In 2001 sales in the commercial sector became higher than in the industrial sector as well. (See Table 8.2.) In 2003 about 1.3 trillion kilowatt-hours went to residential use, 1.0 trillion kilowatt-hours to industrial customers, and 1.1 trillion kilowatt-hours to commercial users.

Consumption of electricity in general is growing because electricity has increasingly taken over the tasks formerly done with coal, natural gas, or human muscle: manufacturing steel, assembling cars, and milking cows. Electricity is being used extensively in technology fields, such as the computer industry. Residential and commercial use of electricity is increasing with new appliances, air conditioners, computers, and many other developing applications.

THE ELECTRIC BILL

The price paid by a consumer for electricity includes the cost of converting energy into electricity from its original form, such as coal, as well as the cost of delivering it. In 2000, according to the EIA's *Annual Energy Review 2003* (2004), consumers paid an average of $20.04 per million Btu for the electric power delivered to their residences, compared to only $5.68 per million Btu for natural gas and $12.01 per million Btu for motor gasoline.

TABLE 8.1

Electricity overview, selected years, 1949–2003

(Billion kilowatthours)

Year	Net generation				Imports[1]		Exports[1]		T & D losses[5] and unaccounted for[6]	End use		
	Electric power sector[2]	Commercial sector[3]	Industrial sector[4]	Total	From Canada	Total	To Canada	Total		Retail sales[7]	Direct use[8]	Total
1949	291	NA	5	296	NA	2	NA	(s)	43	255	NA	255
1950	329	NA	5	334	NA	2	NA	(s)	44	291	NA	291
1955	547	NA	3	550	NA	5	NA	(s)	58	497	NA	497
1960	756	NA	4	759	NA	5	NA	1	76	688	NA	688
1965	1,055	NA	3	1,058	NA	4	NA	4	104	954	NA	954
1970	1,532	NA	3	1,535	NA	6	NA	4	145	1,392	NA	1,392
1971	1,613	NA	3	1,616	NA	7	NA	4	150	1,470	NA	1,470
1972	1,750	NA	3	1,753	NA	10	NA	3	166	1,595	NA	1,595
1973	1,861	NA	3	1,864	NA	17	NA	3	165	1,713	NA	1,713
1974	1,867	NA	3	1,870	NA	15	NA	3	177	1,706	NA	1,706
1975	1,918	NA	3	1,921	NA	11	NA	5	180	1,747	NA	1,747
1976	2,038	NA	3	2,041	NA	11	NA	2	194	1,855	NA	1,855
1977	2,124	NA	3	2,127	NA	20	NA	3	197	1,948	NA	1,948
1978	2,206	NA	3	2,209	NA	21	NA	1	211	2,018	NA	2,018
1979	2,247	NA	3	2,251	NA	23	NA	2	200	2,071	NA	2,071
1980	2,286	NA	3	2,290	NA	25	NA	4	216	2,094	NA	2,094
1981	2,295	NA	3	2,298	NA	36	NA	3	184	2,147	NA	2,147
1982	2,241	NA	3	2,244	NA	33	NA	4	187	2,086	NA	2,086
1983	2,310	NA	3	2,313	NA	39	NA	3	198	2,151	NA	2,151
1984	2,416	NA	3	2,419	NA	42	NA	3	173	2,286	NA	2,286
1985	2,470	NA	3	2,473	NA	46	NA	5	190	2,324	NA	2,324
1986	2,487	NA	3	2,490	NA	41	NA	5	158	2,369	NA	2,369
1987	2,572	NA	3	2,575	NA	52	NA	6	164	2,457	NA	2,457
1988	2,704	NA	3	2,707	NA	39	NA	7	161	2,578	NA	2,578
1989	[2]2,848	4	[4]115	2,967	NA	26	NA	15	223	2,647	108	2,755
1990	2,901	6	131	3,038	16	18	16	16	214	2,713	114	2,827
1991	2,936	6	133	3,074	20	22	2	2	213	2,762	118	2,880
1992	2,934	6	143	3,084	26	28	2	3	224	2,763	122	2,886
1993	3,044	7	146	3,197	29	31	3	4	236	2,861	128	2,989
1994	3,089	8	151	3,248	45	47	1	2	224	2,935	134	3,069
1995	3,194	8	151	3,353	41	43	2	4	235	3,013	144	3,157
1996	3,284	9	151	3,444	42	43	2	3	237	3,101	146	3,247
1997	3,329	9	154	3,492	43	43	7	9	232	3,146	148	3,294
1998	3,457	9	154	3,620	40	40	12	14	221	3,264	161	3,425
1999	3,530	9	156	3,695	43	43	13	14	229	3,312	183	3,495
2000	3,638	8	157	3,802	49	49	13	15	231	3,421	183	3,605
2001	3,580	7	149	3,737	38	39	16	16	P215	3,370	RE174	P3,544
2002	R3,698	R7	R153	P3,858	36	36	13	R14	R241	P3,463	RE178	P3,641
2003	P3,691	P8	P150	P3,848	P29	P30	P24	P24	P179	P3,500	E175	P3,675

TABLE 8.1

Electricity overview, selected years, 1949–2003 [CONTINUED]

(Billion kilowatthours)

[1]Electricity transmitted across U.S. borders with Canada and Mexico.
[2]Electricity-only and combined-heat-and-power (CHP) plants within the NAICS (North American Industry Classification System) 22 category whose primary business is to sell electricity, or electricity and heat, to the public. Through 1988, data are for electric utilities only; beginning in 1989, data are for electric utilities and independent power producers.
[3]Commercial combined-heat-and-power (CHP) and commercial electricity-only plants.
[4]Industrial combined-heat-and-power (CHP) and industrial electricity-only plants. Through 1988, data are for industrial hydroelectric power only.
[5]Transmission and distribution losses (electricity losses that occur between the point of generation and delivery to the customer).
[6]Data collection frame differences and nonsampling error.
[7]Electricity retail sales to ultimate customers by electric utilities and other energy service providers.
[8]Commercial and industrial facility use of onsite net electricity generation; and electricity sales among adjacent or co-located facilities for which revenue information is not available.
R=Revised.
P=Preliminary.
E=Estimate.
NA=Not available.
(s)=Less than 0.5 billion kilowatthours.
Notes: Totals may not equal sum of components due to independent rounding.
Web Pages: For data not shown for 1951–1969, see http://www.eia.doe.gov/emeu/aer/elect.html.
For related information, see http://www.eia.doe.gov/emei/aer/elect.html.

SOURCE: "Table 8.1. Electricity Overview, Selected Years, 1949–2003 (Billion Kilowatthours)," in *Annual Energy Review 2003*, U.S. Department of Energy, Energy Information Administration, Office of Energy Markets and End Use, September 7, 2004, http://www.eia.doe.gov/emeu/aer/pdf/aer.pdf (accessed September 28, 2004)

FIGURE 8.2

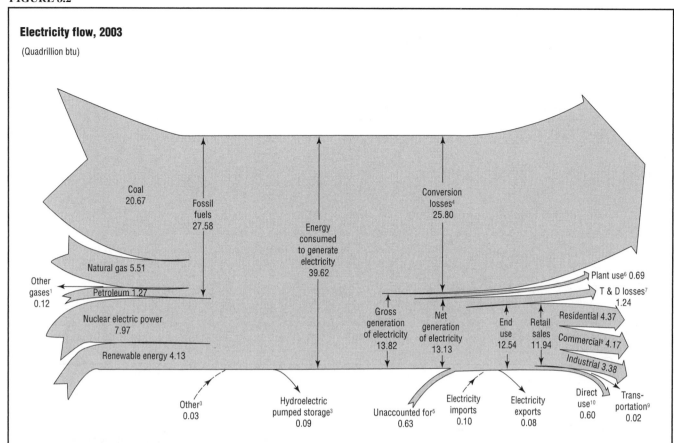

Electricity flow, 2003

(Quadrillion btu)

[1] Blast furnace gas, propane gas, and other manufactured and waste gases derived from fossil fuels.
[2] Batteries, chemicals, hydrogen, pitch, purchased steam, sulfur, and miscellaneous technologies.
[3] Pumped storage facility production minus energy used for pumping.
[4] Approximately two-thirds of all energy used to generate electricity.
[5] Data collection frame differences and nonsampling error.
[6] Electric energy used in the operation of power plants, estimated as 5 percent of gross generation.
[7] Transmission and distribution losses (electricity losses that occur between the point of generation and delivery to the customer) are estimated as 9 percent of gross generation.
[8] Commercial retail sales plus approximately 95 percent of "Other" retail sales.
[9] Approximately 5 percent of "Other" retail sales.
[10] Commercial and industrial facility use of onsite net electricity generation; and electricity sales among adjacent or co-located facilities for which revenue information is not available.
Note: Totals may not equal sum of components due to independent rounding.

SOURCE: "Diagram 5. Electricity Flow, 2003 (Quadrillion Btu)," in *Annual Energy Review 2003,* U.S. Department of Energy, Energy Information Administration, Office of Energy Markets and End Use, September 7, 2004, http://www.eia.doe.gov/emeu/aer/pdf/aer.pdf (accessed September 28, 2004)

The unit cost of electricity is high because of the amount of energy expended in creating the electricity and moving it to the point of use. In 2003, for example, about 39.6 quadrillion Btu of energy were consumed by electric utilities to generate electricity in the United States, but 13.1 quadrillion Btu was the net generation, after accounting for energy used by the power plants themselves. (See Figure 8.2.) Most of the remaining 25.8 quadrillion Btu was lost during the energy conversion process. Additionally, about 1.2 quadrillion Btu is lost during the transmission and distribution process (T & D losses). In the end, for every three units of energy that are converted to create electricity, slightly less than one unit actually reaches the end user.

Between 1960 and 1970, the price of electricity declined, but it began to increase during the 1970s because of the OPEC (Organization of the Petroleum Exporting Countries) oil embargo. (See Figure 8.4.) From the mid-1980s to

2003 the price of electricity, in general, dropped because of the decline in energy resource prices. Prices varied depending upon the location. As Figure 8.5 shows, in 2002 electricity was most expensive in the New England states, New York, New Jersey, Pennsylvania, California, Nevada, Alaska, and Hawaii. According to the EIA's *Electric Power Annual 2002* (December 2003), the average price of electricity sold to the residential sector was 8.5 cents per kilowatt-hour in 2002, while the commercial sector paid 7.9 cents per kilowatt-hour. Industrial users paid less per kilowatt-hour, 4.9 cents in 2002, because the huge amounts of electricity they use allow them to receive volume discounts. The average price for all sectors across the United States in 2002 was 7.2 cents.

DEREGULATION OF ELECTRIC UTILITIES

Regulated for decades as "natural monopolies," much like the railroad and telecommunications industries, elec-

FIGURE 8.3

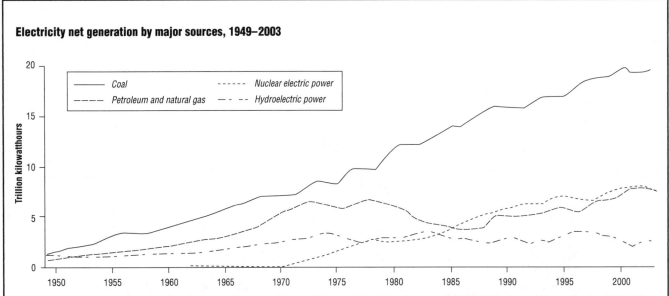

Electricity net generation by major sources, 1949–2003

SOURCE: Adapted from "Figure 8.2a. Electricity Net Generation, Total (All Sectors): By Major Sources, 1949–2003," in *Annual Energy Review 2003*, U.S. Department of Energy, Energy Information Administration, Office of Energy Markets and End Use, September 7, 2004, http://www.eia.doe.gov/emeu/aer/pdf/aer.pdf (accessed September 28, 2004)

tric utilities are in the midst of a radical, highly controversial shift toward unregulated markets and increased competition. In 1978 Congress passed the Public Utilities Regulatory Policy Act (PURPA; PL 95-617), which required that utilities buy electricity from private companies when that would be a lower-cost alternative to building their own power plants. The Energy Policy Act of 1992 (PL 102-486) gave other generators greater access to the market, resulting in a flurry of activity in state and federal legislatures as a host of interest groups debated regulatory, economic, energy, and environmental policies. State public utility commissions conducted proceedings and designed rules related to competition in the electric utility industry.

California was a leader in deregulation activities. In the summer of 2000, however, the state experienced rolling blackouts and electricity bills doubled for many. Fearful of the blackouts and price spikes that afflicted California, by the spring of 2001 most other states had slowed or stopped their efforts to deregulate their electricity markets. At that time, twenty-four states and the District of Columbia had begun deregulation. Then, during an investigation of Enron Corporation, documents were found that showed that Enron electricity traders used strategies that added to electricity costs and congestion on transmission lines for their own profit. These revelations lowered public confidence in power companies in general. As a result of these events and an eroding confidence in deregulation, only seventeen states plus the District of Columbia were actively engaged in restructuring activities as of February 2003, as shown in Figure 8.6 from the

FIGURE 8.4

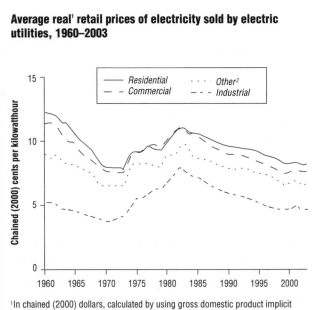

Average real[1] retail prices of electricity sold by electric utilities, 1960–2003

[1]In chained (2000) dollars, calculated by using gross domestic product implicit price deflators.
[2]Public street and highway lighting, other sales to public authorities, sales to railroads and railways, and interdepartmental sales.

SOURCE: Adapted from "Figure 8.10. Average Retail Prices of Electricity: Real, 1960–2003," in *Annual Energy Review 2003*, U.S. Department of Energy, Energy Information Administration, Office of Energy Markets and End Use, September 7, 2004, http://www.eia.doe.gov/emeu/aer/pdf/aer.pdf (accessed September 28, 2004)

Energy Information Administration (EIA). In addition, five states had delayed the restructuring process, and one state (California) suspended its restructuring activities. Restructuring was not active in twenty-seven states.

TABLE 8.2

Electricity end use, selected years, 1949–2003

(Billion kilowatthours)

Year	Retail sales[1]					Direct use[4]	Total
	Residential	Commercial[2]	Industrial[2]	Other[3]	Total		
1949	67	45	123	20	255	NA	255
1950	72	51	146	22	291	NA	291
1955	128	79	260	29	497	NA	497
1960	201	131	324	32	688	NA	688
1965	291	200	429	34	954	NA	954
1970	466	307	571	48	1,392	NA	1,392
1971	500	329	589	51	1,470	NA	1,470
1972	539	359	641	56	1,595	NA	1,595
1973	579	388	686	59	1,713	NA	1,713
1974	578	385	685	58	1,706	NA	1,706
1975	588	403	688	68	1,747	NA	1,747
1976	606	425	754	70	1,855	NA	1,855
1977	645	447	786	71	1,948	NA	1,948
1978	674	461	809	73	2,018	NA	2,018
1979	683	473	842	73	2,071	NA	2,071
1980	717	488	815	74	2,094	NA	2,094
1981	722	514	826	85	2,147	NA	2,147
1982	730	526	745	86	2,086	NA	2,086
1983	751	544	776	80	2,151	NA	2,151
1984	780	583	838	85	2,286	NA	2,286
1985	794	606	837	87	2,324	NA	2,324
1986	819	631	831	89	2,369	NA	2,369
1987	850	660	858	88	2,457	NA	2,457
1988	893	699	896	90	2,578	NA	2,578
1989	906	726	926	90	2,647	108	2,755
1990	924	751	946	92	2,713	114	2,827
1991	955	766	947	94	2,762	118	2,880
1992	936	761	973	93	2,763	122	2,886
1993	995	795	977	95	2,861	128	2,989
1994	1,008	820	1,008	98	2,935	134	3,069
1995	1,043	863	1,013	95	3,013	144	3,157
1996	1,083	887	1,034	98	3,101	146	3,247
1997	1,076	929	1,038	103	3,146	148	3,294
1998	1,130	979	1,051	104	3,264	161	3,425
1999	1,145	1,002	1,058	107	3,312	183	3,495
2000	1,192	1,055	1,064	109	3,421	183	3,605
2001	1,203	1,089	964	114	3,370	RE174	R3,544
2002	R1,267	R1,116	R972	R107	R3,463	RE178	R3,641
2003	P1,280	P1,119	P991	P109	P3,500	E175	P3,675

[1]Electricity retail sales to ultimate customers by electric utilities and, beginning in 1996, other energy service providers.
[2]Retail customers are classified as "Commercial" or "Industrial" based on NAICS (North American Industry Classification System) codes or usage falling within specified limits by rate schedule.
[3]Public street and highway lighting, other sales to public authorities, sales to railroads and railways, and interdepartmental sales.
[4]Commercial and industrial facility use of onsite net electricity generation; and electricity sales among adjacent or co-located facilities for which revenue information is not available.
R=Revised.
P=Preliminary.
E=Estimate.
NA=Not available.
Note: Totals may not equal sum of components due to independent rounding.
Web Pages: For data not shown for 1951–1969, see http://www.eia.doe.gov/emeu/aer/elect.html. For related information, see http://www.eia.doe.gov/fuelelectric.html.

SOURCE: "Table 8.9. Electricity End Use, Selected Years, 1949–2003 (Billion Kilowatthours)," in *Annual Energy Review 2003*, U.S. Department of Energy, Energy Information Administration, Office of Energy Markets and End Use, September 7, 2004, http://www.eia.doe.gov/emeu/aer/pdf/aer.pdf (accessed September 28, 2004)

INTERNATIONAL ELECTRICITY USAGE

World Production

In 2002 approximately 15.3 trillion kilowatt-hours of electricity were generated around the world. (See Figure 8.7.) As reported in the EIA's *Annual Energy Review 2003* (2004), the United States accounted for 25% of this production; China, 10.4%; Japan, 6.5%; and Russia, 5.8%. Figure 8.8 shows net generation of electricity by type and by region of the world.

World Consumption

According to the EIA in *Annual Energy Review 2003* (2004), world total electricity consumption continued to increase, rising from 353.3 quadrillion Btu in 1993 to 411.6 quadrillion Btu in 2002. North America, Central America, and South America accounted for 34% of the world's total consumption in 2002; Asia and Oceania, 28%; Western Europe, 18%; and Eastern Europe and the former USSR, 13%. The Middle East and Africa used 5% and 3%, respectively. (See Figure 8.9.)

FIGURE 8.5

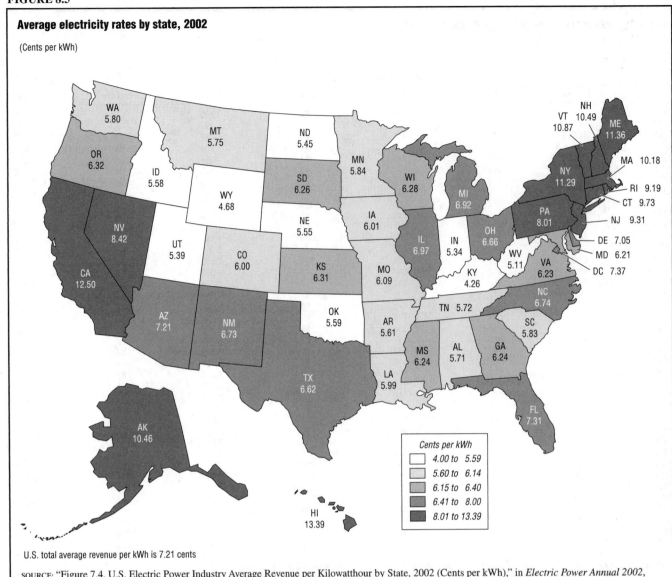

Average electricity rates by state, 2002

(Cents per kWh)

U.S. total average revenue per kWh is 7.21 cents

SOURCE: "Figure 7.4. U.S. Electric Power Industry Average Revenue per Kilowatthour by State, 2002 (Cents per kWh)," in *Electric Power Annual 2002*, U.S. Department of Energy, Energy Information Administration, Office of Coal, Nuclear, Electric and Alternate Fuels, December 2003, http://tonto.eia.doe .gov/FTPROOT/electricity/034802.pdf (accessed November 16, 2004)

FUTURE TRENDS IN THE U.S. ELECTRIC INDUSTRY

In *Annual Energy Outlook 2004* the EIA predicted that from 2002 to 2025, residential U.S. electricity consumption will grow at a rate of 1.4% annually. This compares with 7% growth per year during the 1960s. Several factors led to this decreased growth of electricity consumption, including increased market saturation of electric appliances and improvements in efficiency. Commercial demand is expected to grow by 2.2% per year because of growth in commercial floor space, while industrial demand will likely increase by 1.6% per year as industrial output rises.

Historically, the demand for electricity in the United States has been related to economic growth. This relationship will continue, but electricity use is expected to grow more slowly than the gross domestic product (GDP), a mea-

sure of economic growth. Figure 8.10 shows how electricity sales are related more to economic growth (the GDP) than to population growth. Also, the phrase "five-year moving average" means that each point on the graph is an average of that year and the previous four years' data. This type of averaging is often done to get a better idea of what the real long-term averages would be, without heavy influences on data from cyclical influences, like business or weather cycles.

The issue of electric growth is important and carries financial risks for electric companies. If the industry underestimates future needs for electricity, it could mean power shortages or losses. However, excessive projections of the nation's needs could mean billions of dollars spent on unneeded equipment.

The EIA estimated that the United States will need 356 gigawatts of new generating capacity from 2002 to

FIGURE 8.6

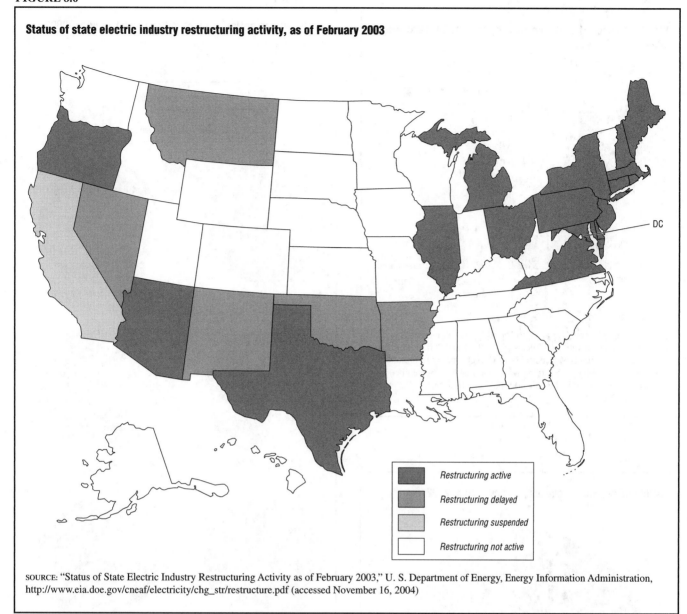

Status of state electric industry restructuring activity, as of February 2003

Legend:
- Restructuring active
- Restructuring delayed
- Restructuring suspended
- Restructuring not active

SOURCE: "Status of State Electric Industry Restructuring Activity as of February 2003," U. S. Department of Energy, Energy Information Administration, http://www.eia.doe.gov/cneaf/electricity/chg_str/restructure.pdf (accessed November 16, 2004)

2025 to meet growing demand for electricity and to replace retiring units, most of it after 2010. From 2002 to 2025, sixty-two gigawatts of capacity will most likely be retired, accounting for nearly all old fossil-fired plants that are not competitive with newer types of fossil-fired plants.

The EIA's *Annual Energy Outlook 2004* projected that electricity prices will decrease by 8% from 2002 to 2008 and will then remain somewhat stable until 2011. From 2011, however, electricity prices are expected to increase gradually by 0.3% per year to 2025.

Continued concerns about acid rain and global warming could result in tightened environmental emission standards, which may have an impact on electrical utility expansion decisions, prices, and supply. Continued advances in solar and wind turbine technology could make renewable sources of electrical power more economical in the future. Some energy experts and environmentalists claim that increased efficiency and conservation efforts are the most sensible alternatives to new construction or to the burning of more fossil fuels in existing plants.

FIGURE 8.7

World net generation of electricity by type, 1980, 1990, and 2002

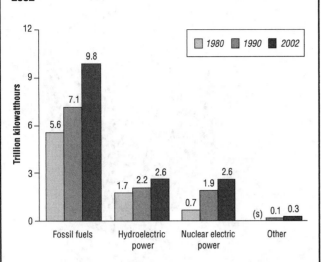

(s)=Less than 0.05 trillion kilowatthours.

SOURCE: Adapted from "Figure 11.16. World Net Generation of Electricity: Net Generation by Type—1980, 1990, and 2002," in *Annual Energy Review 2003,* U.S. Department of Energy, Energy Information Administration, Office of Energy Markets and End Use, September 7, 2004, http://www.eia.doe.gov/emeu/aer/pdf/aer.pdf (accessed September 28, 2004)

FIGURE 8.8

World net generation of electricity by type by region, 2002

(Percent of regional total)

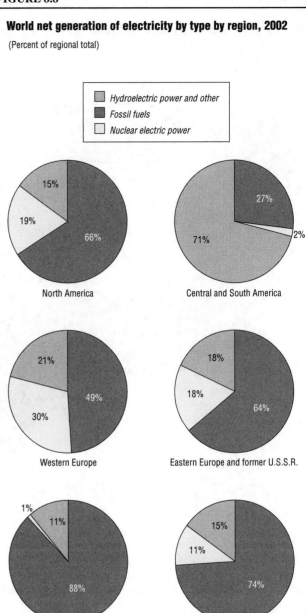

SOURCE: Adapted from "Figure 11.16. World Net Generation of Electricity: Net Generation by Type by Region, 2002 (Percent of Regional Total)," in *Annual Energy Review 2003,* U.S. Department of Energy, Energy Information Administration, Office of Energy Markets and End Use, September 7, 2004, http://www.eia.doe.gov/emeu/aer/pdf/aer.pdf (accessed September 28, 2004)

FIGURE 8.9

World primary energy consumption by region, 2002

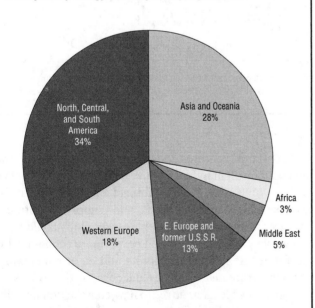

SOURCE: Adapted from "Figure 11.3. World Primary Energy Consumption: Regional Consumption Shares, 2002," in *Annual Energy Review 2003,* U.S. Department of Energy, Energy Information Administration, Office of Energy Markets and End Use, September 7, 2004, http://www.eia.doe.gov/emeu/aer/pdf/aer.pdf (accessed September 28, 2004)

FIGURE 8.10

Population, gross domestic product, and electricity sales, 1965–2025

(5-year moving average annual percent growth)

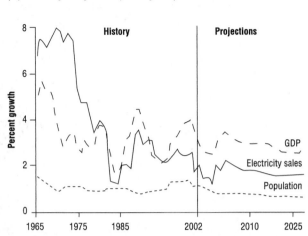

SOURCE: "Figure 67. Population, Gross Domestic Product, and Electricity Sales, 1965–2025 (5-year moving average annual percent growth)," in *Annual Energy Outlook 2004,* U.S. Department of Energy, Energy Information Administration, Office of Integrated Analysis and Forecasting, January 2004, http://tonto.eia.doe.gov/FTPROOT/ forecasting/0383(2004).pdf (accessed November 16, 2004)

CHAPTER 9

ENERGY CONSERVATION

ENERGY CONSERVATION AND EFFICIENCY

Energy conservation is the efficient use of energy, without necessarily curtailing the services that energy provides. Conservation occurs when societies develop efficient technologies that reduce energy needs. Environmental concerns, such as acid rain and the potential for global warming, have increased public awareness about the importance of energy conservation.

Energy efficiency can be measured by two indicators. The first is energy consumption per person (per capita) per year. Annual per person energy consumption in the United States was 215 million Btu in 1949, it topped out at 360 million Btu in 1978 and 1979, dropped to 314 million Btu by 1983, and then slowly rose again until it reached 338 million Btu in 2003, which was 57% above the 1949 rate. (See Figure 9.1.)

The second indicator of efficiency is energy consumption per dollar of gross domestic product (GDP). GDP is the total value of goods and services produced by a nation. When a country grows in its energy efficiency, it uses less energy to produce the same amount of goods and services. In 1949 nearly 19,600 Btu of energy were consumed for each dollar of GDP. (See Figure 9.2.) In 1970 about 17,990 Btu of energy were consumed per dollar, and 9,440 Btu were used per dollar in 2003.

ENERGY CONSERVATION, PUBLIC HEALTH, AND THE ENVIRONMENT

People living in cities with high levels of pollution have higher risks of mortality from certain diseases than those living in less polluted cities. Energy-related emissions generate a vast majority of these polluting chemicals. (Table 4.3 in Chapter 4 shows some air pollutants and their sources.) According to the American Lung Association, air pollution has been related to such diseases as asthma, bronchitis, emphysema, and lung cancer. The

FIGURE 9.1

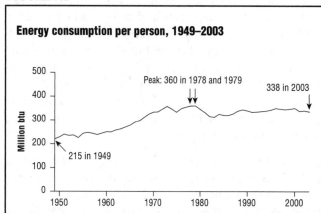

Energy consumption per person, 1949–2003

SOURCE: "Figure 2. Energy Consumption per Person," in *Annual Energy Review 2003,* U.S. Department of Energy, Energy Information Administration, Office of Energy Markets and End Use, September 7, 2004, http://www.eia.doe.gov/emeu/aer/pdf/aer.pdf (accessed September 28, 2004)

FIGURE 9.2

Energy consumption per dollar of gross domestic product, 1950–2003

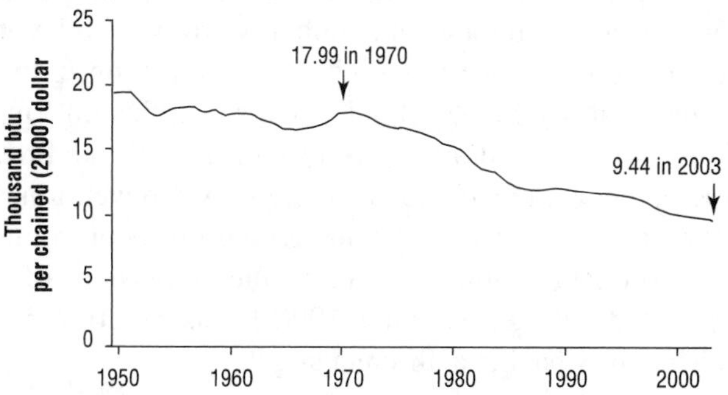

SOURCE: "Figure 3. Energy Use per Dollar of Gross Domestic Product," in *Annual Energy Review 2003,* U.S. Department of Energy, Energy Information Administration, Office of Energy Markets and End Use, September 7, 2004, http://www.eia.doe.gov/emeu/aer/pdf/aer.pdf (accessed September 28, 2004)

group estimated the annual health costs of exposure to the most serious air pollutants to be in the billions. Clean and efficient energy technologies represent a cost-effective investment in public health.

Global warming is long-term climate change—a worldwide temperature increase—caused by the "greenhouse effect." The greenhouse effect is the trapping of heat within the atmosphere by high levels of gases called "greenhouse gases": carbon dioxide, methane, nitrogen oxide, hydrofluorocarbons, sulfur dioxides, and perfluorocarbons. Just as the glass of a greenhouse or the windows of a car trap heat, greenhouse gases keep the earth warmer than if the atmosphere contained only oxygen and nitrogen.

The United Nations' Intergovernmental Panel on Climate Change (IPCC) Examines Global Warming

The United Nations' Intergovernmental Panel on Climate Change (IPCC), a group of 2,000 of the world's leading scientists, was established in 1988 and reported in 1995 that global warming is real, serious, and accelerating. It determined that the most likely cause is primarily humans burning coal, oil, and gasoline, which has increased the amount of carbon dioxide and other greenhouse gases in the earth's atmosphere. Deforestation also contributes by reducing the amount of carbon dioxide absorbed and stored in plants.

The IPCC published its *Third Assessment Report* in 2001. This is the IPCC's most recent report; another is scheduled for publication in 2007. The assessment report noted that scientists now have a clearer understanding of the causes and consequences of global warming because of the intensive climate research and environmental monitoring that has been conducted. They described the effect that global warming will have on weather patterns, water resources, the seasons, ecosystems, and extreme climate events, and they urged governments to move forward quickly with climate change policies.

The Kyoto Protocol

In December 1997 the United Nations convened a 160-nation conference on global warming in Kyoto, Japan, in hopes of producing a new treaty on climate change that would place binding caps on industrial emissions. In its initial draft the treaty, called the Kyoto Protocol to the United Nations Framework Convention on Climate Change (or simply the Kyoto Protocol), bound industrialized nations to reducing their emissions of the six greenhouse gases below 1990 levels by 2012, with each country having a different target.

Under the terms of initial drafts of the Kyoto Protocol, the United States was to cut emissions by 7%, most European nations by 8%, and Japan by 6%. Reductions were to begin by 2008 and to be achieved by 2012. Developing nations were not required to make such pledges.

The United States had proposed a program of voluntary pledges by developing nations, but that section was deleted, as was a tough system of enforcement. Instead, each country was to decide for itself how to achieve its goal. The treaty would provide market-driven tools, such as buying and selling credits, for reducing emissions. It would also set up a Clean Development Fund to help poorer nations with technology to reduce their emissions.

The Kyoto Protocol marked the first time nations made such sweeping pledges to cut emissions, but there were difficulties getting ratification. In the United States, President Bill Clinton signed the protocol but the U.S. Senate did not ratify it. By early 2001 the Kyoto Protocol was near collapse. However, diplomats from 178 nations met in Bonn, Germany, in July 2001 and drafted a compromise in an effort to preserve the global-warming treaty. In October 2001, two thousand delegates from 160 countries began twelve days of talks in Marrakesh, Morocco, in order to complete a final draft of the Kyoto Protocol. President George W. Bush, though, stood firm and rejected the Kyoto Protocol, characterizing it as "fatally flawed" and saying implementation would harm the U.S. economy and unfairly require only the industrial nations to cut emissions.

U.S. feelings on the Kyoto Protocol varied widely. Business leaders believed the treaty went too far, while environmentalists believed the Kyoto standards did not go far enough. Some experts doubted that any action emerging from Kyoto would be sufficient to prevent the doubling of greenhouse gases. Representatives of the oil industry and business community contended that the treaty would spell economic pain for the United States. The fossil-fuel industry and conservative politicians portrayed the protocol as unworkable and too costly to the American economy.

Nevertheless, in November 2004 Russia ratified the treaty. This gave the Kyoto Protocol support from countries whose emissions total 55% of the world's greenhouse gases, which was minimum needed for the treaty to go into effect. By 2004, 126 countries had signed on, including all the European Union members, Japan, and Norway, but the treaty lacked support from the United States and Australia.

Although President Bush rejected the Kyoto Protocol, he promised to address the issue of greenhouse gas emissions and global warming. He proposed studying the problem and funding new technologies to reduce carbon dioxide emissions.

Levels of greenhouse gas emissions for the United States for 1990 and from 1996 to 2002 are shown in Table 9.1 in teragrams of carbon equivalents (Tg). A teragram is a trillion grams. Each gas in the table is reported by its global warming potential (GWP) in Tg, which allows

TABLE 9.1

Trends in greenhouse gas emissions and sinks, 1990 and 1996–2002

Gas/source	1990	1996	1997	1998	1999	2000	2001	2002
CO₂	**5,002.3**	**5,498.5**	**5,577.6**	**5,602.5**	**5,676.3**	**5,859.0**	**5,731.8**	**5,782.4**
Fossil fuel combustion	4,814.7	5,310.1	5,384.0	5,412.4	5,488.8	5,673.6	5,558.8	5,611.0
Iron and steel production	33.3	37.1	38.3	39.2	40.0	41.2	41.4	42.9
Cement manufacture	85.4	68.3	71.9	67.4	64.4	65.7	59.1	54.4
Waste combustion	10.9	17.2	17.8	17.1	17.6	18.0	18.8	18.8
Ammonia production and urea application	19.3	20.3	20.7	21.9	20.6	19.6	16.2	17.7
Lime manufacture	11.2	13.5	13.7	13.9	13.5	13.3	12.8	12.3
Limestone and dolomite use	5.5	7.8	7.2	7.4	8.1	6.0	5.7	5.8
Natural gas flaring	5.8	8.5	7.9	6.6	6.9	5.8	5.4	5.3
Aluminum production	6.3	5.6	5.6	5.8	5.9	5.7	4.1	4.2
Soda ash manufacture and consumption	4.1	4.2	4.4	4.3	4.2	4.2	4.1	4.1
Titanium dioxide production	1.3	1.7	1.8	1.9	1.9	1.9	1.9	2.0
Phosphoric acid production	1.5	1.6	1.5	1.6	1.5	1.4	1.3	1.3
Carbon dioxide consumption	0.9	0.8	0.8	0.9	0.9	1.0	0.8	1.3
Ferroalloys	2.0	2.0	2.0	2.0	1.7	1.3	1.2	
Land-use change and forestry (sink)[1]	(957.9)	(1,055.2)	(821.0)	(705.8)	(675.8)	(690.2)	(689.7)	(690.7)
International bunker fuels[2]	113.9	102.3	109.9	115.1	105.3	101.4	97.9	86.8
Biomass combustion[2]	216.7	244.3	233.2	217.2	222.3	226.8	204.4	207.1
CH₄	**642.7**	**637.0**	**628.8**	**620.1**	**613.1**	**614.4**	**605.1**	**598.1**
Landfills	210.0	208.8	203.4	196.6	197.8	199.3	193.2	193.0
Natural gas system	122.0	127.4	126.1	124.5	120.9	125.7	124.9	121.8
Enteric fermentation	117.9	120.5	18.3	116.7	116.6	115.7	114.3	114.4
Coal mining	81.9	63.2	62.6	62.8	58.9	56.2	55.6	52.2
Manure management	31.0	34.6	36.3	38.8	38.6	38.0	38.8	39.5
Wastewater treatment	24.1	26.9	27.4	27.7	28.2	28.4	28.1	28.7
Petroleum systems	28.9	25.6	25.5	25.0	23.7	23.5	23.5	23.2
Stationary sources	8.2	8.8	7.8	7.2	7.5	7.7	7.2	6.9
Rice cultivation	7.1	7.0	7.5	7.9	8.3	7.5	7.6	6.8
Mobile sources	5.0	4.8	4.7	4.5	4.5	4.4	4.3	4.2
Abandoned coal mines	3.4	6.0	5.6	4.8	4.4	4.4	4.2	4.1
Petrochemical production	1.2	1.6	1.6	1.7	1.7	1.7	1.4	1.5
Iron and Steel production	1.3	1.3	1.3	1.2	1.2	1.2	1.1	1.0
Agricultural residue burning	0.7	0.8	0.8	0.8	0.8	0.8	0.8	0.7
Silicon carbide production	*	*	*	*	*	*	*	*
International bunker fuels[2]	0.2	0.1	0.1	0.2	0.1	0.1	0.1	+ 0.1
N₂O	**393.2**	**436.9**	**436.3**	**432.1**	**428.4**	**425.8**	**417.3**	**415.8**
Agricultural soil management	262.8	288.1	293.2	294.2	292.1	289.7	288.6	287.3
Mobile sources	50.7	60.7	60.3	59.6	58.6	57.4	55.0	52.9
Manure management	16.2	17.0	17.3	17.3	17.4	17.7	18.0	17.8
Nitric acid	17.8	20.7	21.2	20.9	20.1	19.6	15.9	16.7
Human sewage	12.8	14.2	14.4	14.7	15.2	15.3	15.4	15.6
Stationary sources	12.6	13.9	4 14	13.8	13.9	14.4	13.9	14.0
Adipic acid	15.2	17.0	10.3	6.0	5.5	6.0	4.9	5.9
N₂O product usage	4.3	4.5	4.8	4.8	4.8	4.8	4.8	4.8
Field burning of agricultural residues	0.4	0.4	0.4	0.5	0.4	0.5	0.5	0.4
Waste combustion	0.4	0.4	0.4	0.3	0.3	0.4	0.4	0.4
International bunker fuels[2]	1.0	0.9	1.0	1.0	0.9	0.9	0.9	0.8
HFCs, PFCs, and SF6	**90.9**	**114.9**	**121.7**	**135.7**	**134.8**	**139.1**	**129.7**	**138.2**
Substitution of ozone depleting substances	0.3	35.0	46.4	56.5	65.8	75.1	83.4	91.7
HCFC-22 production	35.0	31.1	30.0	40.0	30.4	29.8	19.8	19.8
Electrical transmission and distribution	29.2	24.3	21.7	17.1	16.4	15.9	15.6	14.8
Aluminum production	18.1	12.5	11.0	9.0	8.9	8.9	4.0	5.2
Semiconductor manufacture	2.9	5.5	6.3	7.1	7.2	6.3	4.5	4.4
Magnesium production and Processing	5.4	6.5	6.3	5.8	6.0	3.2	2.5	2.4
Total	**6,129.1**	**6,687.3**	**6,764.4**	**6,790.5**	**6,852.5**	**7,038.3**	**6,883.9**	**6,934.6**
Net emissions sources and sinks	**5,171.3**	**5,632.1**	**5,943.5**	**6,084.7**	**6,176.8**	**6,348.2**	**6,194.1**	**6,243.8**

*Does not exceed 0.05 Tg CO Eq.
[1]Sinks are only included in net emissions total, and are based partially on projected activity data. Parentheses indicate negative values (or sequestration).
[2]Emissions from International Bunker Fuels and Biomass combustion are not included in totals.
Note: Totals may not sum due to independent rounding.

SOURCE: "Table ES-2. Recent Trends in U.S. Greenhouse Gas Emissions and Sinks (TgCO₂Eq.)," in *Inventory of U.S. Greenhouse Gas Emissions and Sinks: 1990–2002*, U.S. Environmental Protection Agency, April 15, 2004, http://yosemite.epa.gov/oar/globalwarming.nsf/UniqueKeyLookup/RAMR5WNMK2/ $File/0 4executivesummary.pdf (accessed November 17, 2004).

these numbers to be compared to one another. Higher numbers indicate greater harm done to the environment by the gas. Conversely, numbers listed in parentheses, such as those for land-use change and forestry, reduce greenhouse gas emissions by the carbon equivalents shown. From 1990 to 2002, there was a net increase of 1,072.5 Tg of greenhouse gas emissions in the United States, which is a 21% increase.

EFFICIENCY IN THE TRANSPORTATION SECTOR

The U.S. transportation system plays a central role in the economy. Highway transportation is dependent on internal combustion engine vehicles fueled almost exclusively by petroleum. The Energy Information Administration (EIA) of the U.S. Department of Energy (DOE) noted in its *Annual Energy Review 2003* (2004) that the transportation sector accounted for 27% of all energy consumed in the United States in 2003. That year, Americans used 26.8 quadrillion Btu of energy for transportation (see Figure 1.9 in Chapter 1), of which petroleum made up 97%. Despite improvements in transportation efficiency in recent decades, the EIA's *International Energy Outlook 2004* noted that the United States consumed 28% of total energy use for transportation in 2001 and is projected to consume 30% in 2025. This transportation share compares to 17% of total energy use for Canada and 23% for Western Europe.

Automotive Efficiency

Policymakers interested in transportation energy conservation have an array of conservation options. (See Table 9.2.) However, not all options are mutually supportive. For example, efforts to promote a freer flow of automobile traffic, such as high-occupancy vehicle (HOV) lanes or free parking for carpools, may sabotage efforts to shift travelers to mass transit or to reduce trip lengths and frequency. Policymakers must consider how the implementation of one strategy will fit into an overall transportation plan.

In the United States the automobile dominates the transportation sector; cars and light-duty vehicles used 63% of all transportation energy in 2004, as reported by the Bureau of Transportation Statistics in *National Transportation Statistics, 2003* (updated September 2004). This report also noted that motor gasoline, which is divided among passenger cars, light and heavy-duty trucks, aircraft, and miscellaneous other modes of transportation, consumed about 67% of the oil used in the United States in 2004. The major growth in fuel use over the past thirty years has been that consumed by pickup trucks, and in more recent years that consumed by vans and sport utility vehicles (SUVs). In 2004 pickup truck, van, and SUV fuel use was approximately 75% of the amount used by automobiles and motorcycles. Meanwhile, automobile fuel use has remained fairly constant because of fuel efficiency increases that have offset the growth in car miles traveled. Boosting truck, van, and SUV efficiency will become increasingly important in holding down oil demand.

THE CORPORATE AVERAGE FUEL ECONOMY (CAFE) STANDARDS. The 1973 OPEC (Organization of the Petroleum Exporting Countries) oil embargo painfully reminded the United States how dependent it had become on foreign sources of fuel. This situation prompted Congress to pass

TABLE 9.2

Transportation conservation options

Improve the technical efficiency of vehicles
1. Higher fuel economy requirements—CAFE standards (R)
2. Reducing congestion: smart highways (E,I), flextime (E,R), better signaling (I), improved maintenance of roadways (I), time of day charges (E), improved air traffic controls (I,R), plus options that reduce vehicular traffic
3. Higher fuel taxes (E)
4. Gas guzzler taxes, or feebate schemes (E)
5. Support for increased R&D (EJ)
6. Inspection and maintenance programs (R)

Increase load factor
1. HOV lanes (I)
2. Forgiven tolls (E), free parking for carpools (E)
3. Higher fuel taxes (E)
4. Higher charges on other vmt trip-dependent factors (E): parking (taxes, restrictions, end of tax treatment as business cost), tolls, etc.

Change to more efficient modes
1. Improvements in transit service
   a. New technologies—maglev, high speed trains (EJ)
   b. Rehabilitation of older systems (I)
   c. Expansion of service—more routes, higher frequency (I)
   d. Other service improvements (I)—dedicated busways, better security, more bus stop shelters, more comfortable vehicles
2. Higher fuel taxes (E)
3. Reduced transit fares through higher US. transit subsidies (E)
4. Higher charges on other vmt/trip-dependent factors for less efficient modes (E)— tolls, parking
5. Shifting urban form to higher density, more mixed use, greater concentration through zoning changes (R), encouragement of "infill" development (E,R,I), public investment in infrastructure (I), etc.

Reduce number or length of trips
1. Shifting urban form to higher density, more mixed use, greater concentration (E,R,I)
2. Promoting working at home or at decentralized facilities (EJ)
3. Higher fuel taxes (E)
4. Higher charges on other vmt/trip-dependent factors (E)

Shift to alternative fuels
1. Fleet requirements for alternative fuel-capable vehicles and actual use of alternative fuels (R)
2. Low-emission/zero emission vehicle (LEV/ZEV) requirements (R)
3. Various promotions (E): CAFE credits, emission credits, tax credits, etc.
4. Higher fuel taxes that do not apply to alternative fuels (E), or subsidies for the alternatives (E)
5. Support for increased R&D (EJ)
6. Public investment—government fleet investments (I)

Freight options
1. RD&D of technology improvements (E,I)

*U.S. transit subsidies, already among the highest in the developed world, may merely promote inefficiencies.

KEY: CAFE = corporate average fuel economy; E = economic incentive; HOV = high-occupancy vehicle; I = public investment; maglev = trains supported by magnetic levitation; R = regulatory action; RD&D = research, development, and demonstration; vmt = vehicle-miles traveled.

SOURCE: "Table 5-1. Transportation Conservation Options," in *Saving Energy in U.S. Transportation,* Office of Technology Assessment, 1994, http://www.wws.princeton.edu/cgibin/byteserv.prl/~ota/disk1/1994/9432/943208.PDF (accessed November 17, 2004)

the 1975 Energy Policy and Conservation Act (PL 94-163), which set the initial Corporate Average Fuel Economy (CAFE) standards. The standards were modified in 1980 with the Automobile Fuel Efficiency Act (PL 96-425).

The CAFE standards required domestic automakers to increase the average mileage of new cars sold to 27.5 miles per gallon (mpg) by 1985. Under CAFE rules, car manufacturers could still sell large, less efficient cars, but to meet the average fuel efficiency rates, they also had to sell smaller, more efficient cars. Automakers that failed to meet each year's CAFE standards were fined. Those that managed to surpass the rates earned credits that they could use in years when they fell below the CAFE requirements. Faced with the CAFE standards, the car companies became more inventive and managed to keep their cars relatively large and roomy while increasing mileage with innovations like electronic fuel injection and using four valves per cylinder.

The CAFE regulations have had a significant effect on fuel efficiency. In the decades since the first oil shock in 1973, the fuel economy of motor vehicles (which includes passenger cars, vans, pickup trucks, SUVs, and trucks) increased from 11.9 mpg in 1973 to 17.0 mpg in 2002. (See Table 9.3.) Greater gains have been made in the economy of passenger cars. In 1974, just after the oil embargo, cars averaged 14.2 mpg (according to the Environmental Protection Agency); in 2003 the average new-car fuel economy was 24.6 mpg. (See Figure 9.3.) The total automobile fleet fuel economy is expected to increase as more fuel-efficient cars enter the market and older, less fuel-efficient autos drop out of the nation's fleet. However, new-car fuel economy has risen only slightly since 1986; since 1988 nearly all gains in automobile efficiency have been offset by increased weight and power in new vehicles.

The Persian Gulf War in 1991 was another strong reminder to the United States of its continuing heavy dependence on foreign oil, prompting some members of Congress to want to raise the CAFE standards to forty-five mpg for cars and thirty-five mpg for light trucks. Those in favor of raising CAFE standards claimed that this would save about 2.8 million barrels of oil per day. They also noted that if cars become even more fuel-efficient in the future, emissions of carbon dioxide would be significantly reduced. The domestic auto industry opposed the bill to raise CAFE standards, and actions to increase automobile efficiency failed in Congress. For model year 2001 passenger cars, the CAFE standard was 27.5 mpg and for light trucks it was 20.7 mpg. The efficiency standard for model year 2005 for light trucks had increased to 21.0 mpg. CAFE standards are set by the National Highway Safety Administration, unless Congress prohibits changes.

SUVs, vans, and pickup trucks make up the "light truck" automotive group, the fastest growing segment of the auto industry. Light trucks accounted for 48% of the U.S. light vehicle market in 2003 (see Figure 9.4) and produced most of the profits of the major auto companies. In 1999 the Environmental Protection Agency (EPA) imposed new regulations tightening emissions standards on cars, minivans, SUVs under 8,500 pounds, and small pickup trucks. This was the first time that SUVs and other light-duty trucks became subject to the same national pollution standards as cars. The standard of an average of

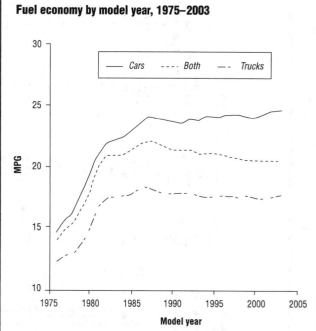

FIGURE 9.3

Fuel economy by model year, 1975–2003

SOURCE: Karl H. Hellman and Robert M. Heavenrich, "Adjusted Fuel Economy by Model Year (Three-Year Moving Average)," in *Light-Duty Automotive Technology and Fuel Economy Trends: 1975 through 2004 Executive Summary,* U.S. Environmental Protection Agency, Advanced Technology Division, Office of Transportation and Air Quality, April 2004, http://www.epa.gov/otaq/cert/mpg/fetrends/420s04002.pdf (accessed November 18, 2004)

0.07 grams per mile for nitrogen oxides came into effect in 2004. Standards for hydrocarbons, nitrogen oxides, carbon monoxide, and particulates were phased in beginning in 1999 and ending in 2008.

The potential for savings from increased fuel economy in large trucks is huge, since their current fuel economy is so much lower than that of automobiles. In 2001 the EIA projected a small increase in fuel efficiency for the heavy-truck fleet and a larger (but still small) increase for the small-truck fleet. If the heavy-truck fleet were to reach a fuel efficiency of ten mpg through technological improvements, projected oil demand would drop by 300,000 barrels per day. Over the past few years, aerodynamically designed trucks have become more common on American roads.

Cheap gasoline prices throughout the 1990s took away the sense of urgency surrounding fuel efficiency, which was demonstrated by the high growth of large vehicle sales. In addition, when the federal fifty-five-mph speed limit law was repealed, many states allowed increased speed limits, lowering fuel efficiency. Carmakers have resisted building highly efficient cars, claiming that government mandates would saddle American motorists with car features they would not want and might not buy. In contrast, the European Commission has proposed an ambitious target of forty-seven mpg for gasoline-driven cars

TABLE 9.3

Motor vehicle mileage, fuel consumption, and fuel rates, selected years, 1949–2002

Year	Passenger cars[1]			Vans, pickup trucks, and sport utility vehicles[2]			Trucks[3]			All motor vehicles[4]		
	Mileage (miles per vehicle)	Fuel consumption (gallons per vehicle)	Fuel rate (miles per gallon)	Mileage (miles per vehicle)	Fuel consumption (gallons per vehicle)	Fuel rate (miles per gallon)	Mileage (miles per vehicle)	Fuel consumption (gallons per vehicle)	Fuel rate (miles per gallon)	Mileage (miles per vehicle)	Fuel consumption (gallons per vehicle)	Fuel rate (miles per gallon)
1949	9,388	627	15.0	5	5	5	9,712	1,080	9.0	9,498	726	13.1
1950	9,060	603	15.0	5	5	5	10,316	1,229	8.4	9,321	725	12.8
1955	9,447	645	14.6	5	5	5	10,576	1,293	8.2	9,661	761	12.7
1960	9,518	668	14.3	5	5	5	10,693	1,333	8.0	9,732	784	12.4
1965	9,603	661	14.5	5	5	5	10,851	1,387	7.8	9,826	787	12.5
1970	9,989	737	13.5	8,676	866	10.0	13,565	2,467	5.5	9,976	830	12.0
1971	10,097	743	13.6	9,082	888	10.2	14,117	2,519	5.6	10,133	839	12.1
1972	10,171	754	13.5	9,534	922	10.3	14,780	2,657	5.6	10,279	857	12.0
1973	9,884	737	13.4	9,779	931	10.5	15,370	2,775	5.5	10,099	850	11.9
1974	9,221	677	13.6	9,452	862	11.0	14,995	2,708	5.5	9,493	788	12.0
1975	9,309	665	14.0	9,829	934	10.5	15,167	2,722	5.6	9,627	790	12.2
1976	9,418	681	13.8	10,127	934	10.8	15,438	2,764	5.6	9,774	806	12.1
1977	9,517	676	14.1	10,607	947	11.2	16,700	3,002	5.6	9,978	814	12.3
1978	9,500	665	14.3	10,968	948	11.6	18,045	3,263	5.5	10,077	816	12.4
1979	9,062	620	14.6	10,802	905	11.9	18,502	3,380	5.5	9,722	776	12.5
1980	8,813	551	16.0	10,437	854	12.2	18,736	3,447	5.4	9,458	712	13.3
1981	8,873	538	16.5	10,244	819	12.5	19,016	3,565	5.3	9,477	697	13.6
1982	9,050	535	16.9	10,276	762	13.5	19,931	3,647	5.5	9,644	686	14.1
1983	9,118	534	17.1	10,497	767	13.7	21,083	3,769	5.6	9,760	686	14.2
1984	9,248	530	17.4	11,151	797	14.0	22,550	3,967	5.7	10,017	691	14.5
1985	9,419	538	17.5	10,506	735	14.3	20,597	3,570	5.8	10,020	685	14.6
1986	9,464	543	17.4	10,764	738	14.6	22,143	3,821	5.8	10,143	692	14.7
1987	9,720	539	18.0	11,114	744	14.9	23,349	3,937	5.9	10,453	694	15.1
1988	9,972	531	18.8	11,465	745	15.4	22,485	3,736	6.0	10,721	688	15.6
1989	R10,157	R533	R19.0	11,676	724	16.1	22,926	3,776	6.1	10,932	688	15.9
1990	10,504	520	20.2	11,902	738	16.1	23,603	3,953	6.0	11,107	677	16.4
1991	10,571	501	21.1	12,245	721	17.0	24,229	4,047	6.0	11,294	669	16.9
1992	10,857	517	21.0	12,381	717	17.3	25,373	4,210	6.0	11,558	683	16.9
1993	10,804	527	20.5	12,430	714	17.4	26,262	4,309	6.1	11,595	693	16.7
1994	10,992	531	20.7	12,156	701	17.3	25,838	4,202	6.1	11,683	698	16.7
1995	11,203	530	21.1	12,018	694	17.3	26,514	4,315	6.1	11,793	700	16.8
1996	11,330	534	21.2	11,811	685	17.2	26,092	4,221	6.2	11,813	700	16.9
1997	11,581	539	21.5	12,115	703	17.2	27,032	4,218	6.4	12,107	711	17.0
1998	11,754	544	21.6	12,173	707	17.2	25,397	4,135	6.1	12,211	721	16.9
1999	11,848	553	21.4	11,957	701	17.0	26,014	4,352	6.0	12,206	732	16.7
2000	11,976	547	21.9	11,672	669	17.4	25,617	4,391	5.8	12,164	720	16.9
2001	R11,831	R534	22.1	R11,204	R636	17.6	R26,602	R4,477	5.9	R11,887	R695	17.1
2002P	12,203	551	22.1	11,365	645	17.6	27,062	4,637	5.8	12,172	715	17.0

TABLE 9.3

Motor vehicle mileage, fuel consumption, and fuel rates, selected years, 1949–2002 [CONTINUED]

[1]Through 1989, includes motorcycles.
[2]Includes a small number of trucks with 2 axles and 4 tires, such as step vans.
[3]Single-unit trucks with 2 axles and 6 or more tires, and combination trucks.
[4]Includes buses and motorcycles, which are not separately displayed.
R = Revised.
P = Preliminary.
Web Pages: For data not shown for 1951–1969, see http://www.eia.doe.gov/emeu/aer/enduse.html.

SOURCE: Table 2.8. Motor Vehicle Mileage, Fuel Consumption, and Fuel Rates, Selected Years, 1949–2002," in *Annual Energy Review 2003*, U.S. Department of Energy, Energy Information Administration, Office of Energy Markets and End Use, September 7, 2004, http://www.eia.doe.gov/emeu/aer/pdf/aer.pdf (accessed September 28, 20

FIGURE 9.4

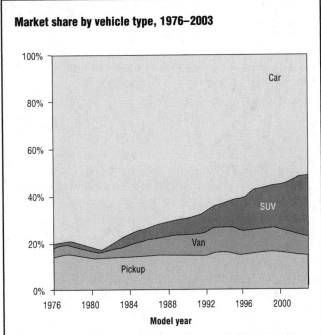

Market share by vehicle type, 1976–2003

SOURCE: Karl H. Hellman and Robert M. Heavenrich, "Sales Fraction by Vehicle Type (Three-Year Moving Average)," in *Light-Duty Automotive Technology and Fuel Economy Trends: 1975 through 2004 Executive Summary,* U.S. Environmental Protection Agency, Advanced Technology Division, Office of Transportation and Air Quality, April 2004, http://www.epa.gov/otaq/cert/mpg/fetrends/420s04002.pdf (accessed November 18, 2004)

(compared to the current average of twenty-nine mpg) and fifty-two mpg for diesel-powered cars by 2005. The general secretary of the European Council of Automotive Research and Development also announced in January 2001 that all ten European auto manufacturers plan to build cars with low carbon dioxide emissions and high mileage by 2008. In 2004 China issued compulsory fuel efficiency standards for its passenger cars, ruling that fuel consumption in Chinese passenger vehicles must decline by 10%. Cars under production will be granted a one-year grace period to meet the standard.

It must be noted that while European countries do not generally legislate fuel efficiency, the cost of gasoline in Europe is more than twice that in the United States. That serves as a powerful incentive to European drivers to buy fuel-efficient vehicles. Many experts also see Europe as having a history of energy consumption not matched by the United States.

Alternative Fuel Vehicles (AFVs)

MANDATING AFVS. Several laws have been passed to encourage or mandate the use of vehicles powered by fuels other than gasoline. The Clean Air Act Amendments of 1990 (PL 101-549) required certain businesses and local governments with fleets of ten or more vehicles in twenty-one metropolitan areas nationwide to phase in alternative

fuel vehicles (AFVs) over time—20% of those fleets had to be AFVs by 1998. While great strides have been made in increasing the use of AFVs, there is no way to determine current compliance with the mandates because reporting and enforcement methods are inadequate.

The Energy Policy Act of 1992 (PL 102-486) was passed in the wake of the 1991 Persian Gulf War to conserve energy and increase the proportion of energy supplied domestically. It required the federal government to purchase 22,500 AFVs by 1995 and increase the percentage of AFV acquisitions from 25% of all acquisitions in 1996 to 75% in 1999 and thereafter. Agency budget cuts and inadequate enforcement have slowed compliance with these regulations. Still, many municipal governments and the U.S. Postal Service have put into operation fleets of natural gas vehicles, such as garbage trucks, transit buses, and postal vans.

NUMBERS AND TYPES OF AFVS. In 1995, 246,855 AFVs were on U.S. roads. Projections for 2004 showed 547,904 AFVs in use. (See Table 9.4.) These totals include vehicles originally manufactured to run on alternative fuels, as well as converted gasoline or diesel vehicles. The manufacture of new AFVs has been steadily increasing.

A number of different types of fuels are used in AFVs:

- Liquefied petroleum gas (LPG) is a mixture of propane and butane. LPG was the most common type of alternative vehicle fuel used in 2004. Thirty-five percent of all AFVs ran on LPG that year.

- Ethanol is ethyl alcohol, a grain alcohol, mixed with gasoline and sold as gasohol. The 85% formulation of gasohol was the second most common AFV fuel, powering 27% of all AFVs in 2004.

- Compressed natural gas (CNG) is natural gas that is stored in pressurized tanks. CNG releases one-tenth the carbon monoxide, hydrocarbon, and nitrogen of gasoline. It was the third most common AFV fuel in 2004, used by 26% of AFVs.

- Electricity, used by 10% of AFVs in 2004, can be used for battery-powered, fuel cell, or hybrid vehicles.

- Methanol is a liquid fuel that can be produced from natural gas, coal, or biomass (plant material, vegetation, or agricultural waste). The 85% formulation of methanol was used by only 0.8% of all AFVs in 2004. Its use is declining.

- Liquefied natural gas (LNG) is natural gas (mostly methane) that has been liquefied by reducing its temperature to -260 degrees Fahrenheit. It was used by only 0.6% of all AFVs in 2004.

- Biodiesels (not listed in Table 9.4) are liquid biofuels made from soybean, rapeseed, or sunflower oil, or

TABLE 9.4

Estimated number of alternative-fueled vehicles in use, by fuel, 1995–2004

Fuel	1995	1996	1997	1998	1999	2000	2001	2002	2003 (Projected)	2004 (Projected)	Average annual growth rate (percent)
Liquefied petroleum gases (LPG)	172,806	175,585	175,679	177,183	178,610	181,994	185,053	187,680	190,438	194,389	1.3
Compressed natural gas (CNG)	50,218	60,144	68,571	78,782	91,267	100,750	111,851	120,839	132,988	143,742	12.4
Liquefied natural gas (LNG)	603	663	813	1,172	1,681	2,090	2,576	2,708	3,030	3,134	20.1
Methanol, 85 percent (M85)[1]	18,319	20,265	21,040	19,648	18,964	10,426	7,827	5,873	4,917	4,592	−14.3
Methanol, neat (M100)	386	172	172	200	198	0	0	0	0	0	0.0
Ethanol, 85 percent (E85)[1,2]	1,527	4,536	9,130	12,788	24,604	87,570	100,303	120,951	133,776	146,195	78.8
Ethanol, 95 percent (E95)[1]	136	361	347	14	14	4	0	0	0	0	0.0
Electricity[3]	2,860	3,280	4,453	5,243	6,964	11,830	17,847	33,047	45,656	55,852	39.1
Non-LPG subtotal	**74,049**	**89,421**	**104,526**	**117,847**	**143,692**	**212,670**	**240,404**	**283,418**	**320,367**	**353,515**	**19.0**
Total	**246,855**	**265,006**	**280,205**	**295,030**	**322,302**	**394,664**	**425,457**	**471,098**	**510,805**	**547,904**	**9.3**

[1]The remaining portion of 85-percent methanol and both ethanol fuels is gasoline.
[2]In 1997, some vehicle manufacturers began including E85-fueling capability in certain model lines of vehicles. For 2002, the EIA estimated that the number of E-85 vehicles that are capable of operating on E85, gasoline, or both, is about 4.1 million. Many of these alternative-fueled vehicles (AFVs) are sold and used as traditional gasoline-powered vehicles. In this table, AFVs in use include only those E85 vehicles believed to be intended for use as AFVs. These are primarily fleet-operated vehicles.
[3]Excludes gasoline-electric hybrids.
Notes: Estimates for 2003, in italics, are based on plans or projections. Estimates for historical years may be revised in future reports if new information becomes available.

SOURCE: "Table 1. Estimated Number of Alternative-Fueled Vehicles in Use in the United States, by Fuel, 1995–2004," in *Alternatives to Traditional Transportation Fuels 2003 Estimated Data*, U.S. Department of Energy, Energy Information Administration, February 2004, http://www.eia.doe.gov/cneaf/alternate/page/datatables/afvtable1_03.xls (accessed November 18, 2004)

TABLE 9.5

Estimated number of alternative-fueled vehicles in use, by state and fuel type, 2002

State	Liquefied petroleum gases	Natural gas	Methanol	Ethanol	Electricity	Total
Alabama	4,289	1,341	0	2,713	636	8,979
Alaska	145	401	0	720	11	1,277
Arizona	1,082	7,243	201	1,583	1,662	11,771
Arkansas	2,199	340	0	300	0	2,839
California	21,537	24,990	4,787	9,517	10,670	71,501
Colorado	5,611	2,694	3	3,491	126	11,925
Connecticut	379	2,762	1	1,849	156	5,147
Delaware	85	489	10	783	11	1,378
District of Columbia	7	1,462	50	1,408	316	3,243
Florida	4,171	4,152	6	7,856	357	16,542
Georgia	4,418	4,484	39	2,076	4,550	15,567
Hawaii	842	0	0	1,467	204	2,513
Idaho	1,581	3,412	0	240	0	5,233
Illinois	5,259	3,120	17	6,916	89	15,401
Indiana	1,426	3,397	0	1,670	91	6,584
Iowa	2,179	18	27	1,903	12	4,139
Kansas	3,565	748	1	1,649	22	5,985
Kentucky	2,214	1,191	0	2,313	0	5,718
Louisiana	1,117	896	3	1,309	0	3,325
Maine	158	77	0	134	21	390
Maryland	2,570	3,634	7	2,901	45	9,157
Massachusetts	249	1,006	36	1,331	78	2,700
Michigan	4,822	991	48	4,840	1,606	12,307
Minnesota	2,162	509	0	3,361	0	6,032
Mississippi	1,193	140	0	543	0	1,876
Missouri	2,642	476	95	3,878	11	7,102
Montana	2,980	268	0	309	0	3,557
Nebraska	4,338	370	0	1,095	11	5,814
Nevada	1,487	3,111	0	973	0	5,571
New Hampshire	718	42	0	169	167	1,096
New Jersey	358	2,723	4	2,681	190	5,956
New Mexico	6,069	1,969	11	2,140	435	10,624
New York	6,213	13,100	88	3,723	9,299	32,423
North Carolina	4,560	559	0	4,539	112	9,770
North Dakota	1,310	155	0	354	0	1,819
Ohio	2,487	2,647	26	4,537	242	9,939
Oklahoma	17,839	3,322	0	1,122	0	22,283
Oregon	3,084	1,034	20	1,528	212	5,878
Pennsylvania	1,107	2,299	108	4,008	89	7,611
Rhode Island	122	331	0	391	0	844
South Carolina	3,047	362	0	4,051	0	7,460
South Dakota	1,374	44	0	384	0	1,802
Tennessee	2,623	763	0	3,068	200	6,654
Texas	39,279	9,961	162	6,706	82	56,190
Utah	3,227	1,961	8	1,966	0	7,162
Vermont	366	5	0	199	178	748
Virginia	927	4,735	7	3,740	1,086	10,495
Washington	4,397	1,925	73	2,760	11	9,166
West Virginia	39	378	0	595	0	1,012
Wisconsin	1,459	1,207	35	3,075	37	5,813
Wyoming	2,368	303	0	87	22	2,780
U.S. total	**187,680**	**123,547**	**5,873**	**120,951**	**33,047**	**471,098**

Notes: Natural gas includes compressed natural gas (CNG) and liquefied natural gas (LNG). Methanol includes M85 and M100. Ethanol includes E85 and E95. Excludes gasoline-electric hybrids. Totals may not equal sum of components due to independent rounding.

SOURCE: "Table 4. Estimated Number of Alternative-Fueled Vehicles in Use, by State and Fuel Type, 2002," in *Alternatives to Traditional Transportation Fuels 2003 Estimated Data*, U.S. Department of Energy, Energy Information Administration, February 2004, http://www.eia.doe.gov/cneaf/alternate/page/datatables/afvtable4_03.xls (accessed November 18, 2004)

from animal tallow. They can also be made from agricultural products, such as rice hulls.

The largest numbers of AFVs are located in California, Texas, New York, and Oklahoma. Together, these four states account for 39% of the estimated number of AFVs in use in 2002. (See Table 9.5.) Transit buses are one type of heavy-duty vehicle that has seen much AFV activity. In 2003 in the United States, 9,278 alternative fuel buses were in use. Most of these were transit buses, rather than school buses, intercity buses, or trolleys ("Table 35: Reported Number of Onroad AFV Buses in Use, by Bus Type, Fuel Type and Configuration, 2003," Energy Information Administration, http://www.eia.doe.gov/ [accessed January 10, 2005]).

ALTERNATIVE FUEL AND THE MARKETPLACE. AFVs cannot become a viable transportation option unless a fuel supply is readily available. Ideally, an infrastructure for

TABLE 9.6

Alternative fuel station counts, by state and fuel type, as of November 22, 2004

State	CNG	E85	LPG	ELEC	BD	HY	LNG	Totals by state
Alabama	8	0	60	0	0	0	0	68
Alaska	0	0	9	0	0	0	0	9
Arizona	28	2	76	26	3	1	8	144
Arkansas	4	0	60	0	0	0	0	64
California	192	2	299	519	14	7	35	1,068
Colorado	25	10	74	4	11	0	0	124
Connecticut	14	0	26	4	1	0	0	45
Delaware	3	0	5	0	3	0	0	11
District of Columbia	1	0	0	0	0	1	0	2
Florida	26	2	109	6	3	0	0	146
Georgia	23	0	47	1	3	0	0	74
Hawaii	0	0	6	4	3	0	0	13
Idaho	8	1	31	0	2	0	1	43
Illinois	10	17	79	0	4	0	0	110
Indiana	12	0	45	0	11	0	0	68
Iowa	0	9	35	0	1	0	0	45
Kansas	3	2	52	0	4	0	0	61
Kentucky	0	4	21	0	0	0	0	25
Louisiana	12	0	25	0	0	0	0	37
Maine	0	0	14	0	3	0	0	17
Maryland	16	3	24	1	5	0	0	49
Massachusetts	12	0	35	33	1	0	0	81
Michigan	15	3	104	2	12	1	0	137
Minnesota	4	100	47	0	2	0	0	153
Mississippi	0	0	26	0	1	0	0	27
Missouri	6	6	104	0	1	0	0	117
Montana	3	2	38	0	2	0	1	46
Nebraska	1	5	26	0	1	0	0	33
Nevada	17	0	27	0	8	1	0	53
New Hampshire	1	0	23	11	4	0	0	39
New Jersey	18	1	20	0	0	0	0	39
New Mexico	9	3	62	0	2	0	0	76
New York	38	0	61	9	0	0	0	108
North Carolina	8	2	61	0	24	0	0	95
North Dakota	4	1	18	0	0	0	0	23
Ohio	17	1	74	0	8	0	0	10
Oklahoma	54	1	81	1	0	0	0	137
Oregon	16	0	47	4	5	0	0	72
Pennsylvania	47	0	87	0	2	0	1	137
Rhode Island	6	0	6	1	0	0	0	13
South Carolina	4	1	42	0	1	0	0	48
South Dakota	0	6	25	0	0	0	0	31
Tennessee	2	1	49	0	3	0	0	55
Texas	39	0	787	6	2	0	6	840
Utah	65	3	32	0	1	0	0	101
Vermont	1	0	14	9	0	0	0	24
Virginia	16	2	39	0	7	0	2	66
Washington	21	1	74	4	14	0	0	114
West Virginia	3	0	8	0	0	0	0	11
Wisconsin	20	5	59	0	1	0	0	85
Wyoming	12	1	36	0	3	0	0	52
Totals by fuel:	**844**	**197**	**3,209**	**645**	**176**	**11**	**54**	**5,136**

Notes: CNG is Compressed Natural Gas, E85-85% is Ethanol, LPG is Propane, ELEC is Electric, BD is Biodiesel, HY is Hydrogen and LNG is Liquefied Natural Gas

SOURCE: "Alternative Fueling Station Counts by State and Fuel Type," U.S. Department of Energy, Alternative Fuels Data Center, November 22, 2004, http://www.eere.energy.gov/afdc/infrastructure/station_counts.html (accessed November 22, 2004)

supplying alternative fuels would be developed simultaneously with the AFVs. Table 9.6 shows the types and numbers of alternative fuel stations available in each state. As of November 22, 2004, there were 5,136 alternative refueling sites in the United States. In 2004 privately owned vehicles used 71% of alternative fuel, state and local vehicles used 26%, and federal vehicles used 3%. (See Table 9.7.)

Many state policies and programs encourage the use of alternative fuels. California, for example, required that 10% of vehicles for sale in the state by 2003 be zero-emission vehicles, such as hybrid electric vehicles. This has caused vehicle manufacturers to expedite vehicle research and development. In fact, electric vehicles are already selling in California, and some rental car agencies now offer them to customers at prices only slightly higher than those for gasoline-powered cars.

Chrysler Corporation stopped making natural gas–powered vehicles after the 1997 model year because it had

TABLE 9.7

Estimated consumption of alternative transportation fuels, by vehicle ownership, 2000, 2002, and 2004

(Thousand gasoline-equivalent gallons)

Fuel	2000				2002				2004			
	Federal	State/local	Private	Total	Federal	State/local	Private	Total	Federal	State/local	Private	Total
Liquefied petroleum gases (LPG)	458	19,730	192,388	212,576	213	20,097	202,833	223,143	164	19,277	222,927	242,36
Compressed natural gas (CNG)	6,294	34,061	46,390	86,745	6,142	52,482	62,046	120,670	7,449	73,217	78,798	159,464
Liquefied natural gas (LNG)	101	6,031	1,127	7,259	124	7,797	1,461	9,382	146	9,035	1,687	10,868
Methanol, 85 percent (M85)*	2	193	390	585	2	103	232	337	1	82	174	257
Methanol, neat (M100)	0	0	0	0	0	0	0	0	0	0	0	0
Ethanol, 85 percent (E85)*	2,661	5,482	3,928	12,071	3,495	9,215	5,073	17,783	4,331	12,713	5,361	22,40
Ethanol, 95 percent (E95)*	0	13	0	13	0	0	0	0	0	0	0	0
Electricity	639	595	1,824	3,058	191	1,165	5,918	7,274	291	1,461	10,084	11,836
Total	**10,155**	**66,105**	**246,047**	**322,307**	**10,167**	**90,859**	**277,563**	**378,589**	**12,382**	**115,785**	**319,031**	**447,198**

*The remaining portion of 85-percent methanol and both ethanol fuels is gasoline. Consumption data include the gasoline portion of the fuel.

Notes: Fuel quantities are expressed in a common base unit of gasoline-equivalent gallons to allow comparisons of different fuel types. Gasoline-equivalent gallons do not represent gasoline displacement. Gasoline equivalent is computed by dividing the lower heating value of the alternative fuel by the lower heating value of gasoline and multiplying this result by the alternative fuel consumption value. Lower heating value refers to the Btu content per unit of fuel excluding the heat produced by condensation of water vapor in the fuel. Totals may not equal sum of components due to independent rounding. Estimates for 2004, in italics, are based on plans or projections. Estimates for historical years may be revised in future reports if new information becomes available.

SOURCE: "Table 13. Estimated Consumption of Alternative Transportation Fuels in the United States, by Vehicle Ownership, 2000, 2002, and 2004," in *Alternatives to Traditional Transportation Fuels 2003 Estimated Data*, U.S. Department of Energy, Energy Information Administration, February 2004, http://www.eia.doe.gov/cneaf/alternate/page/datatables/afvtable13_03.xls (accessed November 18, 2004)

lost money on the vehicles, selling only 4,000 after production began in 1992. General Motors, which had suspended sales of natural-gas vehicles in 1994, resumed sales in 1997. Ford began selling some natural-gas versions of its cars and trucks in 1995. Commercial fleets, not retail customers, are the main buyers of natural-gas vehicles.

Market success of alternative fuels and AFVs depends upon public acceptance. People are accustomed to using gasoline as their main transportation fuel and it is readily available. As federal and state requirements for alternative fuels increase, so will the fuels' visibility and acceptance by the general public.

ELECTRIC CARS: PROMISE AND REALITY. In the early days of the automobile, electric cars outnumbered internal-combustion vehicles. With the introduction of technology for producing low-cost gasoline, however, electric vehicles fell out of favor. But as cities became choked with air pollution, the idea of an efficient electric car emerged. To make it acceptable to the public, however, several considerations had to be addressed: How many miles could an electric car be driven before needing to be recharged? How light would the vehicle need to be? And could the electric car keep up with the speed and driving conditions of busy freeways and highways?

Electric vehicles (EVs) are of three types: battery-powered; fuel cell; and hybrids, which are powered by both an electric motor and a small conventional engine. EV1, a two-seater by General Motors (GM), was the first commercially available electric car. In 1999 GM introduced its second-generation EV1, the Gen II. It used a lead-acid battery pack and had a driving range of approximately ninety-five miles. The Gen II was also offered with an optional nickel-metal hydride battery pack, which increased its range to 130 miles. However, the California Air Resources Board relaxed automobile-emissions requirements and GM subsequently found that it could no longer market the EV1 effectively. When leases on the cars ran out in 2003, GM began reclaiming the cars.

Fuel cell electric vehicles use an electrochemical process that converts a fuel's energy into usable electricity. Some experts think that in the future vehicles driven by fuel cells could replace vehicles with combustion engines. Fuel cells produce very little sulfur and nitrogen dioxide and generate less than half the carbon dioxide of internal-combustion engines. Rather than needing to be recharged, they are simply refueled. Hydrogen, natural gas, methanol, and gasoline can all be used with a fuel cell.

DaimlerChrysler's Mercedes-Benz division produced the first prototype fuel-cell car. The NECAR4 produces zero emissions and runs on liquid hydrogen. The hydrogen must be kept cold at all times, which makes the design impractical for widespread use. However, the company plans to replace the NECAR4 with the NECARX,

which will run on methanol and is expected to be more practical. The NECAR4 prototype travels 280 miles on a full eleven-gallon tank. It was unveiled in 1999, and was still being road tested in late 2004. Ecostar, an alliance between Ford, DaimlerChrysler, and Ballard Power Systems, is also working on developing new fuel cells to power vehicles.

Hybrid cars have both an electric motor and a small internal-combustion engine. A sophisticated computer system automatically shifts from the electric motor to the gas engine, as needed, for optimum driving. The electric motor is recharged while the car is driving and braking. Because the gasoline engine does only part of the work, these cars get very good fuel economy. The engines are also designed for ultralow emissions.

As of 2002 two hybrid passenger cars were introduced in the United States: the Toyota Prius, a sedan with front and back seating, and the two-passenger Honda Insight. Both cars were sold in Japan for several years before being introduced to the U.S. market. In the 2005 model year, Ford offered the first hybrid SUV, the Escape, which won the 2004 North American Truck of the Year award. The vehicle is reported to get thirty-five mpg with city driving, traveling about four hundred miles on a fifteen-gallon tank.

Air Travel Efficiency

Flying carries an environmental price, as it is a very energy-intensive form of transportation. In much of the industrialized world, air travel is replacing more energy-efficient rail or bus travel. The U.S. Department of Transportation Bureau of Transportation Statistics reported that despite a rise in the fuel efficiency of jet engines, jet fuel consumption rose 75% between 1995 and 2002, from 2.1 billion liters to 3.7 billion liters.

Jet fuel consumption can affect global warming. Airplanes spew nearly four million tons of nitrogen oxide into the air, much of it while cruising in the tropospheric zone five to seven miles above the earth, where ozone is formed. (See Figure 9.5.) The EPA estimated that air traffic accounts for about 3% of all global greenhouse warming. The Intergovernmental Panel on Climate Change (IPCC) for the United Nations noted that emissions deposited directly into the atmosphere do greater harm than those released at the earth's surface.

In 1996 Pratt and Whitney, a designer and manufacturer of high-performance engines, announced plans to introduce a radical new engine design that would be cleaner, more efficient, quieter, and more reliable than conventional designs. The new engine underwent detailed design in 2002, was tested in 2004, and is expected to take its first flight in 2006. The engine would reduce emissions by 40% and exceed noise restriction standards that took

FIGURE 9.5

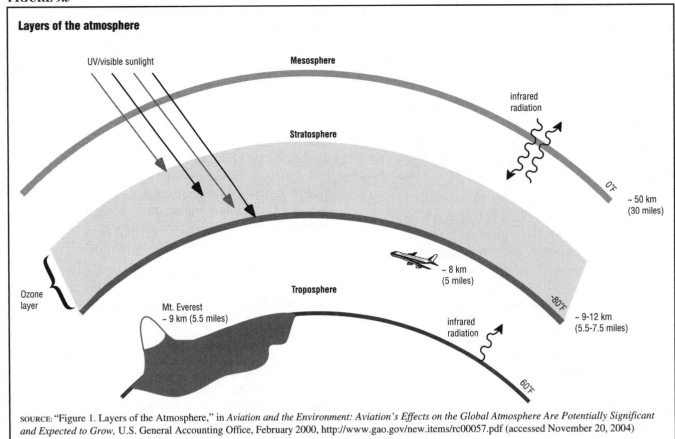

Layers of the atmosphere

UV/visible sunlight

Mesosphere

infrared radiation

Stratosphere

~ 50 km (30 miles)

0°F

Ozone layer

Troposphere

~ 8 km (5 miles)

Mt. Everest ~ 9 km (5.5 miles)

infrared radiation

-80°F

~ 9-12 km (5.5-7.5 miles)

60°F

SOURCE: "Figure 1. Layers of the Atmosphere," in *Aviation and the Environment: Aviation's Effects on the Global Atmosphere Are Potentially Significant and Expected to Grow,* U.S. General Accounting Office, February 2000, http://www.gao.gov/new.items/rc00057.pdf (accessed November 20, 2004)

effect in 2000. The engine is designed for use on single-aisle planes carrying 120 to 180 passengers, such as the Airbus A320.

Although each generation of airplane engines gets cleaner and more fuel-efficient, there are also other engines in the airline industry—those in the trucks, cars, and carts that service airplane fleets. Electric utility companies, including the Edison Electric Institute and the Electric Power Research Institute, launched a program in 1993 to electrify airports. By converting terminal transport buses, food trucks, and baggage-handling carts to electricity, airports could reduce air pollution considerably. As of December 2004, only a few U.S. airports and airlines were operating significant numbers of electric ground support equipment and the associated electric charging stations. However, airport electrification implementation and research were ongoing.

CONSERVATION IN THE RESIDENTIAL AND COMMERCIAL SECTORS

Total building energy use in the United States has increased because there are increasing numbers of people, households, and offices. However, energy use per unit area (commercial) or per person (residential) has roughly stabilized over the past ten to twelve years because of a variety of efficiency improvements. The sources of energy

in buildings have changed dramatically. Use of fuel oil has dropped, and natural gas has largely made up the difference. At the same time, other energy demands have risen. Electronic office equipment, such as computers, fax machines, printers, and copiers, has sharply increased electricity loads in commercial buildings. The Energy Information Administration (EIA) reported in its *Annual Energy Review 2003* that energy use in the residential and commercial sectors accounted for an increasing share of total U.S. energy consumption: 29% in 1950, 33% in 1970, and 40% in 2003. (See Figure 1.9 in Chapter 1.)

Building Efficiency

There are several potential areas for research and development in energy conservation in buildings, which is where energy is used in the residential and commercial sectors. Among the techniques useful in reducing energy loads are advanced window designs, daylighting (letting light in from the outside by using high windows, skylights, and atria in the center of large buildings), solar water heating, landscaping, and tree planting. Energy conservation efforts in buildings have been substantial since the early 1980s.

Residential energy consumption can be reduced by introducing more efficient new housing and appliances, improving energy efficiency in existing housing, and building more multiple-family units. Residential energy con-

FIGURE 9.6

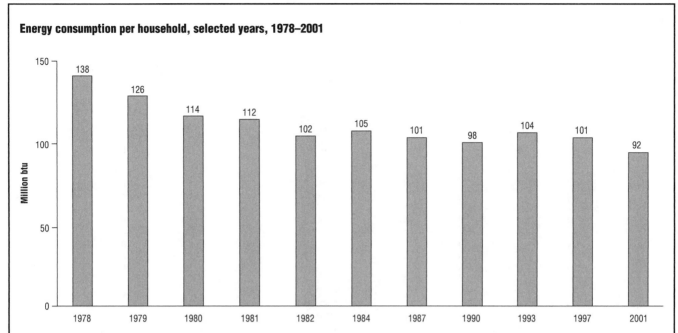

Energy consumption per household, selected years, 1978–2001

Note: For years not shown, there are no data available. Data for 1978 through 1984 are for April of the year shown through March of the following year; data for 1987, 1990, 1993, 1997 and 2001 are for calendar year.

SOURCE: Adapted from "Figure 2.4. Household Energy Consumption: Consumption per Household, Selected Years, 1978–2001," in *Annual Energy Review 2003,* U.S. Department of Energy, Energy Information Administration, Office of Energy Markets and End Use, September 7, 2004, http://www.eia.doe .gov/emeu/aer/pdf/aer.pdf (accessed September 28, 2004)

sumption is reduced as well when people migrate to the South and West, where the combined use of heating and cooling is generally lower than in other parts of the country.

In the residential sector, the amount of energy used in newer homes, particularly those built since 1980, is dramatically less than that used in older homes. The largest share of energy savings is the result of better construction, higher quality insulation, and more energy-efficient windows and doors. The Office of Technology Assessment reported in May 1992 (in *Building Energy Efficiency*) that roughly one-fourth the energy used to heat and cool buildings is lost through poor insulation and poorly insulated windows. Before the 1973 energy crisis, 70% of new windows sold were single-glazed (had only a single pane of glass). By 1990, because of changes in building codes and public interest, 80% of windows sold were double-glazed with double insulating ability, cutting energy loss in half. Double-glazed windows have two panes of glass sandwiched together with a small space in between. In addition, the glass may be specially treated or the space between the panes may be filled with a gas, either of which increases the insulating effectiveness of the window.

Energy consumption per household has remained fairly steady since 1982, as technology gains have been offset by an increase in the size of new homes and more demand for energy services. (See Figure 9.6.) As in the residential sector, improved technology has helped to

slow the growth in commercial building energy use. Commercial buildings constructed after 1980 use considerably less energy than those built in the early part of the 1900s.

Home Appliance Efficiency

Overall, the number of households in the United States is increasing, which is increasing the demand for energy-intensive products and services like air-conditioning and appliances. The Energy Information Administration (EIA) reported that residential energy use accounted for nearly 22% of the total national energy use in 2003. (See Figure 1.9 in Chapter 1.) For household energy consumption in 2001 (the most recent data compiled by the EIA), space heating used 47% of the total energy consumed, down from 51% in 1997; appliances 30%, up from 27%; water heating 17%, down from 19%; and air conditioners 6%, up from 4%.

The percentages of households with electric appliances has increased steadily over the last couple of decades. (See Figure 9.7.) By 2001, 99% of American homes had color televisions, 86% had microwave ovens, 79% had clothes washers, and 56% had personal computers.

In 1987 Congress passed the National Appliance Energy Conservation Act (NAECA; PL 100-12), which gave the Department of Energy the authority to formulate minimum efficiency requirements for thirteen classes of consumer products. It could also revise and update those standards as technologies and economic conditions

FIGURE 9.7

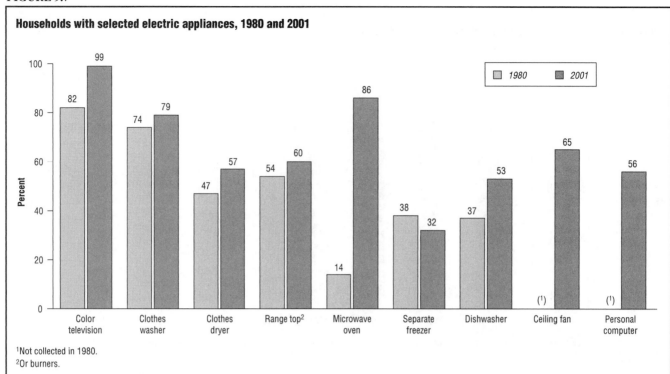

Households with selected electric appliances, 1980 and 2001

[Legend: 1980, 2001]

Color television: 82, 99
Clothes washer: 74, 79
Clothes dryer: 47, 57
Range top[2]: 54, 60
Microwave oven: 14, 86
Separate freezer: 38, 32
Dishwasher: 37, 53
Ceiling fan: (1), 65
Personal computer: (1), 56

[1]Not collected in 1980.
[2]Or burners.

SOURCE: Adapted from "Figure 2.6. Households with Selected Appliances and Types of Main Heating Fuel: Households with Selected Electric Applicances, 1980 and 2001," in *Annual Energy Review 2003*, U.S. Department of Energy, Energy Information Administration, Office of Energy Markets and End Use, September 7, 2004, http://www.eia.doe.gov/emeu/aer/pdf/aer.pdf (accessed September 28, 2004)

changed. Table 9.8 shows the products affected and the years in which appliance efficiency standards were established or revised for each, as well as the future effective dates of standards.

Energy efficiency has increased for all major household appliances but most dramatically for refrigerators and freezers. Since 1972 new refrigerators and freezers have more than tripled in energy efficiency because of better insulation, motors, compressors, and accessories such as automatic defrost ("Good Stuff? A Behind-the-Scenes Guide to the Things We Buy: Appliances," WorldWatch Institute, http://www.worldwatch.org/pubs/goodstuff/appliances/this.href [accessed January 14, 2005]). These improvements have been accomplished at relatively low cost to manufacturers. In addition, efficiency labels for consumers are now required, which makes purchasing efficient models easier.

According the U.S. Department of Energy, by the early 2000s, air conditioners and heat pumps, another major group of appliances, had shown a 30–50% improvement in energy efficiency since the mid-1970s. Although this improvement in energy efficiency was less than the improvements in refrigerators and freezers, it is significant because these appliances are large energy users.

Water heaters and furnaces improved efficiency between 5% and 20% from the mid-1970s to the early 2000s. However, the technological improvements in these appliances are relatively costly compared to the overall price of the product. This means that the more energy-conserving models have a higher retail price, which discourages many consumers from purchasing efficient models, even though the more efficient models may save money in the long run. In addition, many purchases of water heaters and furnaces are made by builders, who have little incentive to pay more for the most efficient models, or by homeowners in emergency situations, when fast availability and installation seem much more important than energy efficiency.

Nonetheless, consumers are sometimes willing to purchase more expensive, energy-efficient models of air conditioners, refrigerators, and lights if the devices can save them enough money in the long run on their electricity bills to offset the higher purchase costs. According to a U.S. General Accounting Office study (*Energy Conservation: Efforts Promoting More Efficient Use*, Washington, DC, 1992), consumers will purchase such devices if the "payback period" is two years or less.

In addition to concerns about efficiency, appliance makers, especially those who make refrigerators and air-conditioning systems, are striving to develop alternative cooling techniques as substitutes for chlorofluorocarbons (CFCs), which are ozone-damaging chemicals that can no longer be legally sold in the United States. Current technol-

TABLE 9.8

Effective dates of appliance efficiency standards, 1988–2007

Product	1988	1990	1992	1993	1994	1995	2000	2001	2003	2004	2005	2006	2007
Clothes dryers	X				X								
Clothes washers	X				X					X			X
Dishwashers	X				X								
Refrigerators and freezers		X		X				X					
Kitchen ranges and ovens		X											
Room air conditioners		X					X						
Direct heating equipment		X											
Fluorescent lamp ballasts		X									X		
Water heaters		X								X			
Pool heaters		X											
Central air conditioners and heat pumps			X									X	
Furnaces													
Central (>45,000 Btu per hour)			X										
Small (<45,000 Btu per hour)			X										
Mobile home		X											
Boilers			X										
Fluorescent lamps, 8 foot					X								
Fluorescent lamps, 2 and 4 foot (U tube)						X							
Commercial water-cooled air conditioners										X			
Commercial natural gas furnaces										X			
Commercial natural gas water heaters										X			

SOURCE: "Table 2. Effective Dates of Appliance Efficiency Standards, 1988–2007," in *Annual Energy Outlook 2002,* U.S. Department of Energy, Energy Information Administration, Office of Integrated Analysis and Forecasting, December 2001, http://tonto.eia.doe.gov/FTPROOT/forecasting/0383(2002).pdf (accessed November 22, 2004)

ogy is temporarily substituting CFCs with somewhat less dangerous HCFCs (hydrochlorofluorocarbons). In Europe, refrigeration units using other substances, such as propane "greenfreeze" technology, are rapidly replacing HCFCs.

Lawn and Garden Equipment

In 1994 the Environmental Protection Agency (EPA) reported that as much as 10% of the nation's air pollution was generated by gasoline-powered lawn and garden equipment, including lawn mowers, chain saws, and golf carts. Former EPA administrator Carol Browner estimated that Americans use eighty-nine million pieces of such equipment, with lawn mowers alone accounting for 5% of the nation's air pollution.

Under Browner, the agency established engine label and warranty requirements, exhaust emissions standards, and test procedures, requiring that engine makers meet the new requirements by 1996. Effective that year, new products offered for sale were equipped with improved carburetion systems, and additional standards were scheduled for subsequent years. Agency officials reported in 2000 that the new regulations had reduced smog-forming hydrocarbon emissions by 32% and that additional standards to be phased in between 2001 and 2007 were projected to reduce hydrocarbon emissions an additional 10%.

INTERNATIONAL COMPARISONS OF CONSERVATION EFFORTS

Compared to other industrialized countries, the United States is lagging behind in energy efficiency and conserva-

tion efforts. According to the Energy Information Administration's *International Energy Annual 2002* (2004) figures for energy consumption per dollar of GDP in 2002, the United States consumed 10,575 Btu per dollar of GDP compared to 5,998 for France, 5,269 for Germany, and 3,876 for Japan. This means that these other countries are consuming less energy per dollar of the output of goods and services; they have lower energy intensities. Put simply, they are making products more efficiently. If fuel prices increase in the future, the United States may face economic challenges from more efficient countries.

As the world's largest producer of greenhouse gases, U.S. per capita emissions are also significantly higher than in other industrialized countries. For instance, the *International Energy Annual 2002* compared countries regarding the metric tons of carbon dioxide they produce per thousand dollars (in 1995 U.S. dollars) of GDP; this figure is termed the carbon intensity of the country. The figures for 2002 showed that the carbon intensity for the United States was substantially higher than for most other industrialized nations. For example, the carbon intensity for the U.S. was 0.62, while it was 0.31 for Germany, 0.22 for France, and 0.21 for Japan.

FUTURE TRENDS IN CONSERVATION

From the Energy Information Administration (EIA)

The EIA's *Annual Energy Outlook 2004* projected U.S. total energy consumption to increase from 97.7 quadrillion Btu to 136.5 quadrillion Btu between 2002 and 2025, an average annual increase of 1.5%, even with

FIGURE 9.8

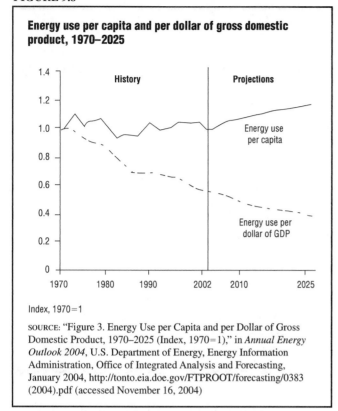

Energy use per capita and per dollar of gross domestic product, 1970–2025

Index, 1970=1

SOURCE: "Figure 3. Energy Use per Capita and per Dollar of Gross Domestic Product, 1970–2025 (Index, 1970=1)," in *Annual Energy Outlook 2004*, U.S. Department of Energy, Energy Information Administration, Office of Integrated Analysis and Forecasting, January 2004, http://tonto.eia.doe.gov/FTPROOT/forecasting/0383 (2004).pdf (accessed November 16, 2004)

FIGURE 9.9

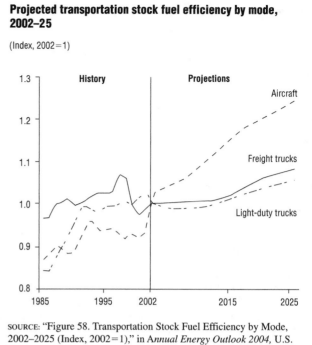

Projected transportation stock fuel efficiency by mode, 2002–25

(Index, 2002=1)

SOURCE: "Figure 58. Transportation Stock Fuel Efficiency by Mode, 2002–2025 (Index, 2002=1)," in *Annual Energy Outlook 2004*, U.S. Department of Energy, Energy Information Administration, Office of Integrated Analysis and Forecasting, January 2004, http://tonto.eia.doe.gov/FTPROOT/forecasting/0383(2004).pdf (accessed November 16, 2004)

efficiency standards for new energy-using equipment in buildings and for motors taken into consideration. Residential and industrial energy demands are expected to be below this average annual increase, at 1.0% per year and 1.3% per year, respectively. However, commercial and transportation energy demands will exceed the average, at 1.7% per year and 1.9% per year, respectively.

Per capita energy use is expected to increase slightly by 0.7% per year from 2002 through 2025. (See Figure 9.8.) Projections anticipate homes will be larger, but that electricity will be used more efficiently. Although annual personal highway and air travel will increase, efficiency improvements will offset much of the increase. Energy use per dollar of GDP (energy intensity) is also projected to decrease as efficiency gains are made.

The EIA projected that transportation fuel efficiency will improve more slowly from 2002 through 2025 than it did during the 1980s. (See Figure 9.9.) Light-duty vehicle efficiency is projected to improve by 6%. For light-duty vehicles, including cars, the EIA projected that any gains in efficiency will be offset by consumer preferences for larger, more powerful vehicles. By 2025 fuel efficiency for large trucks is expected to increase about 9%, while a larger 22% gain is expected for new aircraft.

The EIA predicted that the market for Alternative Fuel Vehicles (AFVs) will grow as a result of federal and state mandates. By 2025 AFVs are projected to account for 3.9 million vehicle sales, or 19% of total light-duty vehicle

sales. According to the *Annual Energy Outlook 2004*, alternative fuels could replace about 166,500 barrels of oil per day by 2025, or 2.1% of all light-vehicle consumption.

From the Bush Administration

The National Energy Policy Plan (NEPP) of the Department of Energy Organization Act of 1977 (PL 95-91) requires the president to submit to Congress a national energy policy plan every two years. This plan includes energy objectives and strategies, as well as projections of energy supply, demand, and prices. In May 2001 the Bush administration submitted the most recent plan, *National Energy Policy: Report of the National Energy Policy Development Group*. In late 2003, a measure to enact significant changes in national energy policy was defeated in Congress.

The 2001 report focused on long-term strategies to provide reliable energy and a clean environment. To do so, the report proposed modernizing the nation's infrastructure, which involved repairing and expanding its outdated network of electric generators, transmission lines, pipelines, and refineries. The report also proposed a modernization of conservation by increasing funding for renewable energy and energy efficiency research, creating tax incentives for purchasing hybrid vehicles, extending the "Energy Star" efficiency program, and reviewing the CAFE standards for transportation.

The report recognized America's dependence on a narrow range of energy sources and supported research into clean coal technologies, the increased use of nuclear power, and the use of new technologies in oil and natural gas exploration. It also proposed opening a small part of the Arctic National Wildlife Refuge to environmentally regulated exploration. Finally, the report promoted further improvements in emission control and the export of environmentally friendly technologies to other parts of the world.

IMPORTANT NAMES AND ADDRESSES

American Gas Association
400 N. Capitol St. NW, Suite 450
Washington, DC 20001
(202) 824-7000
FAX: (202) 824-7115
URL: http://www.aga.org

American Petroleum Institute
1220 L St. NW
Washington, DC 20005-4070
(202) 682-8000
FAX: (202) 962-4730
E-mail: pr@api.org
URL: http://www.api.org

American Wind Energy Association
122 C St. NW, Suite 380
Washington, DC 20001
(202) 383-2500
FAX: (202) 383-2505
E-mail: windmail@awea.org
URL: http://www.awea.org

Committee on Resources
U.S. House of Representatives
1324 Longworth House Office Bldg.
Washington, DC 20515
(202) 225-2761
E-mail: resources.committee@mail.house.
gov
URL: http://resourcescommittee.house.gov

Council on Environmental Quality
722 Jackson Pl. NW
Washington, DC 20503
(202) 395-5750
FAX: (202) 456-6546
URL: http://www.whitehouse.gov/ceq

Edison Electric Institute
701 Pennsylvania Ave. NW
Washington, DC 20004-2696
(202) 508-5000
FAX: (202) 508-5759
E-mail: news@eei.org
URL: http://www.eei.org

Electric Power Research Institute
3412 Hillview Ave.
PO Box 10412
Palo Alto, CA 94304
E-mail: askepri@epri.com
URL: http://www.epri.com

Energy Information Administration
U.S. Department of Energy
1000 Independence Ave. SW
Washington, DC 20585
(202) 586-8800
FAX: (202) 586-0727
E-mail: infoctr@eia.doe.gov
URL: http://www.eia.doe.gov

Environmental Defense
257 Park Ave. S
New York, NY 10010
(212) 505-2100
FAX: (212) 505-2375
E-mail: members@environmentaldefense.
org
URL: http://www.environmentaldefense.org

Friends of the Earth
1717 Massachusetts Ave. NW, Suite 600
Washington, DC 20036-2002
Toll-free: 1-877-843-8687
FAX: (202) 783-0444
E-mail: foe@foe.org
URL: http://www.foe.org

Greenpeace USA
702 H St. NW
Washington, DC 20001
(202) 462-1177
Toll-free: 1-800-326-0959
FAX: (202) 462-4507
E-mail: info@wdc.greenpeace.org
URL: http://www.greenpeaceusa.org

National Mining Association
101 Constitution Ave. NW, Suite 500 East
Washington, DC 20001-2133

(202) 463-2600
FAX: (202) 463-2666
E-mail: craulston@nma.org
URL: http://www.nma.org

Natural Gas Supply Association
805 15th St. NW, Suite 510
Washington, DC 20005
(202) 326-9300
FAX: (202) 326-9330
URL: http://www.ngsa.org

Natural Resources Defense Council
40 West 20th St.
New York, NY 10011
(212) 727-2700
FAX: (212) 727-1773
E-mail: nrdcinfo@nrdc.org
URL: http://www.nrdc.org

Nuclear Energy Institute
1776 I St. NW, Suite 400
Washington, DC 20006-3708
(202) 739-8000
FAX: (202) 785-4019
E-mail: webmasterp@nei.org
URL: http://www.nei.org

Public Citizen
1600 20th St. NW
Washington, DC 20009
(202) 588-1000
FAX: (202) 588-7799
E-mail: member@citizen.org
URL: http://www.citizen.org

Sierra Club
85 2nd St., 2nd Fl.
San Francisco, CA 94105
(415) 977-5500
FAX: (415) 977-5799
E-mail: information@sierraclub.org
URL: http://www.sierraclub.org

Solid Waste Association of North America
P.O. Box 7219

Silver Spring, MD 20907-7219
Toll-free: 1-800-467-9262
FAX: (301) 589-7068
E-mail: info@swana.org
URL: http://www.swana.org

Union of Concerned Scientists
2 Brattle Sq.
Cambridge, MA 02238-9105
(617) 547-5552
FAX: (617) 864-9405
E-mail: ucs@ucsusa.org
URL: http://www.ucsusa.org

U.S. Bureau of Land Management
1849 C St.
Washington, DC 20240
(202) 452-5125
FAX: (202) 452-5124
URL: http://www.blm.gov

U.S. Department of Energy
1000 Independence Ave. SW
Washington, DC 20585
(202) 586-5575
Toll-free: 1-800-DIAL-DOE
FAX: (202) 586-4403
E-mail: the.secretary@hq.doe.gov
URL: http://www.energy.gov

U.S. Environmental Protection Agency
Ariel Rios Bldg.
1200 Pennsylvania Ave. NW
Washington, DC 20460
(202) 272-0167
URL: http://www.epa.gov

U.S. Nuclear Regulatory Commission
Washington, DC 20555
(301) 415-8200
Toll-free: 1-800-368-5642

FAX: (301) 415-2234
E-mail: opa@nrc.gov
URL: http://www.nrc.gov

U.S. Senate Committee on Energy and Natural Resources
364 Dirksen Senate Bldg.
Washington, DC 20510
(202) 224-4971
FAX: (202) 224-6163
URL: http://energy.senate.gov

Worldwatch Institute
1776 Massachusetts Ave. NW
Washington, DC 20036-1904
(202) 452-1999
FAX: (202) 296-7365
E-mail: worldwatch@worldwatch.org
URL: http://www.worldwatch.org

RESOURCES

The Energy Information Administration (EIA) of the U.S. Department of Energy (DOE) is the major source of energy statistics in the United States. It publishes weekly, monthly, and yearly statistical collections on most types of energy; they are available in libraries and online at http://www.eia.doe.gov. The *Annual Energy Review 2003* (2004) provided a complete statistical overview, while the *Annual Energy Outlook 2004* (2004) projected these findings into the future. The EIA's *International Energy Annual 2002* (2004) presents a statistical overview of the world energy situation, while the *International Energy Outlook 2004* (2004) projects these findings into the future. The *U.S. Crude Oil, Natural Gas, and Natural Gas Liquids Reserves 2002 Annual Report* (2003) discusses reserves of coal, oil, and gas. The DOE's Office of Policy provides information on the National Energy Policy Plan in *Report of the National Energy Policy Development Group* (2001).

The DOE/EIA also provide *Natural Gas Annual 2002* (2004) and *Electric Power Annual 2002* (2003). The DOE makes available information on the Waste Isolation Pilot Project, the development of alternative vehicles and fuels,

renewable energy sources, and electric industry restructuring.

The U.S. Department of Transportation's Bureau of Transportation Statistics provides transportation information in its *Transportation Statistics Annual Report 2004* (2004).

The U.S. Environmental Protection Agency (EPA) maintains an Internet home page (http://www.epa.gov/radiation/yucca/index.html) for the Yucca Mountain Repository. Information from that site was extremely useful in the updating of this book. The EPA also provides *Light-Duty Automotive Technology and Fuel Economy Trends 1975–2004* (2004) and *Inventory of U.S. Greenhouse Gas Emissions and Sinks: 1990–2002* (2004).

The U.S. Nuclear Regulatory Commission (NRC) and the United Nations are also important sources of information. The NRC provides the document *NRC—Regulator of Nuclear Safety* and *Information Digest 2004–2005 Edition* (2004). The United Nations Environmental Programme (UNEP) provides *Climate Change Information Kit* (2001), a document based on the work of the Intergovernmental Panel on Climate Change (IPCC).

INDEX

Page references in italics refer to photographs. References with the letter t following them indicate the presence of a table. The letter f indicates a figure. If more than one table or figure appears on a particular page, the exact item number for the table or figure being referenced is provided.

A

Acid rain
    coal and, 62, 64
    electricity and, 123
    future of coal and, 68
Active solar energy systems, 97
AD (associated-dissolved) natural gas, 50
advanced high-temperature reactor (AHTR), 79
AFVs. *See* Alternative Fuel Vehicles
Agricultural wastes, burning of, 88
AHTR (advanced high-temperature reactor), 79
Air conditioner efficiency, 142
Air pollution
    air pollutants, health risks, and contributing sources, 64t
    Clean Air Act of 1990, 64–65
    coal plants and, 63
    electric vehicles to reduce, 140
    energy conservation and, 127–128
    lawn and garden equipment and, 143
    See also Emissions
Air travel
    atmosphere layers and air travel, 140f
    efficiency, 139–140
Airplane engines, 139–140
Alaska
    crude oil production in, 31, 34
    crude oil proved reserves, 103
    crude oil, proved reserves of, by selected states and state subdivisions, 103t
    drilling in Arctic National Wildlife Refuge, 4, 28, 108
    Exxon Valdez oil spill, 39–40
    North Slope fields, 42, 50, 53
    refinery in, 25

Alaska National Interest Lands Conservation Act, 108
Alternative energy. *See* Renewable energy
Alternative fuel
    marketplace and, 136–137, 139
    station counts, by state and fuel type, 137t
Alternative Fuel Vehicles (AFVs)
    alternative transportation fuels, by vehicle ownership, 138t
    electric cars, 139
    in use, by fuel, 135t
    in use, by state and fuel type, 136t
    laws, 134
    numbers/types of, 134, 136–137, 139
    projections, 144
American Community Survey (Census Bureau), 45
American Lung Association, 127–128
American Wind Energy Association, 96
Amperes (amps), 115
Animals
    drilling in Arctic National Wildlife Refuge, 108
    work by, 1
Annual Energy Outlook (Energy Information Administration), 13–14
Annual Energy Outlook 2003 (Energy Information Administration), 14
Annual Energy Outlook 2004 (Energy Information Administration)
    coal production forecasts, 68
    domestic crude oil production, 31, 34
    electricity, future trends, 122–123
    electricity price projections, 14
    energy conservation future trends, 143–144
    gas industry future trends, 50, 53
    new refineries in U.S., 25
    nuclear energy, 74
    renewable energy, future trends, 101
    wind energy production, 96
Annual Energy Review 2000 (Energy Information Administration), 38
Annual Energy Review 2003 (Energy Information Administration)

coal consumption, 60
coal emissions control, 65
coal reserves, 108
crude oil imports, 31
domestic oil production, 25
electricity, cost of, 116
electricity usage, 115–116
energy consumption in residential/commercial sectors, 140
geothermal energy production, 95
international electricity usage, 121
international energy usage, 9, 12
international nuclear energy production, 74–75
natural gas exports, 50
natural gas production, 41
transportation sector efficiency, 130
U.S. oil imports, 30
wood burning statistics, 88
world coal production, 68
Anthracite, 58
ANWR. *See* Arctic National Wildlife Refuge
Appliances
    effective dates of appliance efficiency standards, 143t
    home appliance efficiency, 141–143
    households with selected electric appliances, 142f
Arctic National Wildlife Refuge (ANWR)
    Bush administration's proposal for, 28, 145
    drilling in, 4, 108
Ash, 92
Asia
    collapse of economies in, 3
    wind energy development, 97
    wind turbines in, 95
    See also China; Japan
Associated-dissolved (AD) natural gas, 50
Atmosphere
    air travel efficiency, 139
    coal plants and, 62, 65
    global warming, 128
    layers of, 140f
    nuclear safety problems, 79
    solar energy, 97

electricity net generation by major
  sources, 120 (f8.3)
electricity production with, 115, 116
international production, 74–75, 77–78
low-level waste compacts, 83f
low-level waste disposal site, 82f
nuclear chain reaction, 73f
nuclear fuel cycle, 74 (f5.5)
nuclear generating units, 76t
nuclear share of electricity net
  generation, 71f
operable nuclear generating units, 74
  (f5.6)
operating nuclear reactors, 75f
overview of, 71
power plants, modern, 78–79
pressurized water reactor, 72 (f5.2)
process of, 71–73
radioactive waste disposal, 81–85
radioactive waste, types of, 80
safety problems, 79–80
sources of radiation, 72 (f5.3)
uranium mill tailings disposal sites, 80f
U.S. production of, 4
usage by sector, 9
world net nuclear electric power
  generation, 77t
Yucca Mountain waste site, 84f
Nuclear Regulatory Commission (NRC)
  number of nuclear power plants in
    operation, 75
  Peach Bottom nuclear power plant and,
    79
  radioactive waste disposal and, 81
Nuclear Waste Policy Act, 81, 82
"Number of Uranium Mills and Plants
  Producing Uranium Concentrate"
  (Energy Information Administration),
  112

O

Ocean power
  ocean thermal energy conversion, 101
  tidal power, 100
  wave energy, 100–101
Ocean thermal energy conversion (OTEC),
  101
OECD (Organization for Economic
  Cooperation and Development)
  member country oil consumption, 28–29
  uranium reserves, 114
Office of Technology Assessment, 141
Offshore drilling
  domestic oil production, 25, 28, 31
  natural gas, 41–42
  number of rigs in operation, 105
  world production increase, 113
Oil
  crude oil distillation, 24f
  crude oil production/crude oil well
    productivity, by geographic location, 30
    (f2.7)
  crude oil production/crude oil well
    productivity, by site, 30 (f2.6)
  crude oil-producing countries, selected,
    34f
  dependency, concerns about, 30–31

domestic consumption, 28
domestic production, 25–28
fuel prices, retail motor gasoline/on-
  highway diesel, 39t
imports/exports, 29–30
imports, future trends in, 14
industry, loss of jobs in, 25
petroleum flow, 29f
petroleum production, 28f
petroleum production, selected years,
  26t–27t
petroleum products, shares by sector, 31f
petroleum reserves, strategic, 34
petroleum supply/consumption/imports,
  37 (f2.11)
petroleum traps, 21f
prices, 34, 37–39
prices in twentieth century, 2–3
projected supply/consumption, 31, 34
quest for, 21
refineries, numbers/capacity, 23, 25
refinery capacity and utilization, selected
  years, 23t
refining methods, 22–23
renewable energy history and, 87
rotary drilling system, 22f
Strategic Petroleum Reserve, end-of-year
  stocks in, 37 (f2.12)
Strategic Petroleum Reserve stocks as
  days of net imports, 38f
transportation, environmental concerns
  about, 39–40
types of, 22
U.S. production of, 4
uses for, 22
world crude oil production, 32t–33t
world petroleum consumption, 35t–36 t
world petroleum consumption, selected
  OECD consumers, 37 (f2.10)
world production/consumption, 28–29
See also Crude oil
Oil & Gas Journal (Pennwell Publishing
  Company), 112, 113t
Oil embargo
  CAFE standards after, 130
  electricity cost and, 119
Oil Pollution Act of 1990, 40
Oil spill, Exxon Valdez, 39–40
Oil wells
  by geographic location, 30 (f2.7)
  cost to drill, 105, 108
  costs of crude oil and natural gas wells
    drilled, 109t–110t
  crude oil and natural gas exploration and
    development by region, major U.S.
    energy companies' expenditures for,
    111t
  crude oil and natural gas exploratory and
    development wells, selected years,
    106t–107t
  crude oil and natural gas rotary rigs in
    operation by type, 108f
  exploratory, 105
  productivity, by site, 30 (f2.6)
Oklahoma, 136
Onshore rigs, 105

OPEC. See Organization of Petroleum
  Exporting Countries
Organic material. See Biomass energy
Organization for Economic Cooperation and
  Development (OECD)
  member country oil consumption, 28–29
  uranium reserves, 114
Organization of Petroleum Exporting
  Countries (OPEC)
  electricity cost and, 119
  oil embargo, 130
  oil prices and, 2–3, 38
  refining in, 25
Oscillating water column (OWC) systems,
  100
OTEC (ocean thermal energy conversion),
  101
Oxygen, 88
Ozone
  air travel and, 139
  atmosphere layers and air travel, 140f
  chlorofluorocarbons and, 142
  coal plants and, 64

P

Palos Verdes Landfill (Rolling Hills Estates,
  CA), 92
Passenger cars
  CAFE standards and, 131
  motor vehicle mileage, fuel consumption,
    and fuel rates, 132t–133t
  See also Automobiles
Passive solar energy systems, 97
Peach Bottom nuclear power plant (PA), 79
Pennsylvania, 79
Pennwell Publishing Company, 112, 113t
Permian Basin, 103
Persian Gulf War, 3, 131
Petroleum
  consumption, future trends, 14
  critical events related to, 2f
  effect on coal industry, 57
  electricity net generation by major
    sources, 120 (f8.3)
  electricity production with, 115, 116
  flow, 29f
  liquefied petroleum gas, 134
  production, 28f
  production, selected years, 26t–27t
  products, shares by sector, 31f
  supply/consumption/imports, 37 (f2.11)
  transportation sector consumption of, 130
  traps, 21f
  usage by sector, 9
  world consumption, 35t–36t
  world consumption, selected OECD
    consumers, 37 (f2.10)
Philippines, 95
Photovoltaic (PV) cell solar energy system
  described, 98
  for rural areas, 99
  future development trends, 99–100
Pickup trucks
  CAFE standards for, 131
  fuel use by, 130
  market share by vehicle type, 134f

Uranium Mill Tailing Radiation Control
    Act, 81
Uranium 235 (U-235), 72–73
U.S. Bureau of Transportation Statistics,
    130
*U.S. Crude Oil, Natural Gas, and Natural
    Gas Liquids Reserves 2002 Annual
    Report* (Energy Information
    Administration), 104
U.S. Department of Energy (DOE)
    alternative energy research/development,
        87
    energy reserves estimates by, 103
    home appliance efficiency, 141–142
    hydrogen fuel initiative, 101
    natural gas reserves, 42
    See also Energy Information
        Administration
U.S. Department of the Interior (DOI), 42,
    108
U.S. Department of Transportation Bureau
    of Transportation Statistics, 139
U.S. Fish and Wildlife Service, 108
U.S. General Accounting Office, 25, 142
U.S. Geological Survey, 113
U.S. Postal Service, 134
"U.S. Uranium Reserves by Forward-Cost
    Category, Year-End 1993–2003" (Energy
    Information Administration), 112

V

Vans
    CAFE standards for, 131
    fuel use by, 130
    market share by vehicle type, 134*f*
    motor vehicle mileage, fuel
        consumption, and fuel rates, 132*t*–133*t*

W

War on Terror, 7, 31

Waste. *See* Radioactive waste
Waste-to-energy plants, 91–92
Water. *See* Geothermal energy; Hydropower
Water heaters, 142
Watt, 115
Watt, James, 115
Wattage, 115
Wave energy power plants, 100–101
Wells
    cost to drill, 105, 108
    costs of crude oil and natural gas wells
        drilled, 109*t*–110*t*
    crude oil and natural gas exploration and
        development by region, major U.S.
        energy companies' expenditures for,
        111*t*
    crude oil and natural gas exploratory and
        development wells, selected years,
        106*t*–107*t*
    crude oil and natural gas rotary rigs in
        operation by type, 108*f*
    exploratory wells, 105
    natural gas, 41
    natural gas average productivity, 44
        (*f*3.5)
    number of natural gas-producing, 44
        (*f*3.4)
    wellhead prices, natural gas, 50*f*
WGC (World Geothermal Congress), 95
Wildlife
    drilling in Arctic National Wildlife
        Refuge, 108
    wind turbines and birds, 97
Wind energy
    alternative to fossil fuels, 87
    domestic energy production by wind
        turbines, 95–97
    in general, 95
    international development of, 97

nonhydroelectric renewable electricity
    generation by energy source, 102
    (*f*6.11)
production projections, 101
pros/cons of, 97
usage, 88
wind capacity map, 96*f*
Wind-energy-production tax credit, 97
"Wind Power Set to Become World's
    Leading Energy Source" (Brown), 97
"Wind Powering America" initiative, 96–97
Wind turbines, 95–97
Windmill, 95
Windows, 141
Wood
    burning for biomass energy, 88
    ethanol from, 91
World. *See* International
World Geothermal Congress (WGC), 95
World Oil (Gulf Publishing Company), 112,
    113*t*
"World Oil Markets" (Energy Information
    Administration), 113
World Petroleum Assessment 2000 (U.S.
    Geological Survey), 113
"World Undiscovered Assessment Results
    Summary" (U.S. Geological Survey),
    113
World Watch Institute, 142
Wyoming, 112

Y

Yellowcake, 73
Yom Kippur War, 1–2
Yucca Mountain (NV)
    map of, 84*f*
    waste site, 82–83, 85